Alex Graves

# Supervised Sequence Labelling with Recurrent Neural Networks

## Studies in Computational Intelligence, Volume 385

**Editor-in-Chief**
Prof. Janusz Kacprzyk
Systems Research Institute
Polish Academy of Sciences
ul. Newelska 6
01-447 Warsaw
Poland
*E-mail:* kacprzyk@ibspan.waw.pl

---

Further volumes of this series can be found on our homepage: springer.com

Vol. 362. Pascal Bouvry, Horacio González-Vélez, and Joanna Kolodziej (Eds.)
*Intelligent Decision Systems in Large-Scale Distributed Environments*, 2011
ISBN 978-3-642-21270-3

Vol. 363. Kishan G. Mehrotra, Chilukuri Mohan, Jae C. Oh, Pramod K. Varshney, and Moonis Ali (Eds.)
*Developing Concepts in Applied Intelligence*, 2011
ISBN 978-3-642-21331-1

Vol. 364. Roger Lee (Ed.)
*Computer and Information Science*, 2011
ISBN 978-3-642-21377-9

Vol. 365. Roger Lee (Ed.)
*Computers, Networks, Systems, and Industrial Engineering 2011*, 2011
ISBN 978-3-642-21374-8

Vol. 366. Mario Köppen, Gerald Schaefer, and Ajith Abraham (Eds.)
*Intelligent Computational Optimization in Engineering*, 2011
ISBN 978-3-642-21704-3

Vol. 367. Gabriel Luque and Enrique Alba
*Parallel Genetic Algorithms*, 2011
ISBN 978-3-642-22083-8

Vol. 368. Roger Lee (Ed.)
*Software Engineering, Artificial Intelligence, Networking and Parallel/Distributed Computing 2011*, 2011
ISBN 978-3-642-22287-0

Vol. 369. Dominik Ryżko, Piotr Gawrysiak, Henryk Rybinski, and Marzena Kryszkiewicz (Eds.)
*Emerging Intelligent Technologies in Industry*, 2011
ISBN 978-3-642-22731-8

Vol. 370. Alexander Mehler, Kai-Uwe Kühnberger, Henning Lobin, Harald Lüngen, Angelika Storrer, and Andreas Witt (Eds.)
*Modeling, Learning, and Processing of Text Technological Data Structures*, 2011
ISBN 978-3-642-22612-0

Vol. 371. Leonid Perlovsky, Ross Deming, and Roman Ilin (Eds.)
*Emotional Cognitive Neural Algorithms with Engineering Applications*, 2011
ISBN 978-3-642-22829-2

Vol. 372. António E. Ruano and Annamária R. Várkonyi-Kóczy (Eds.)
*New Advances in Intelligent Signal Processing*, 2011
ISBN 978-3-642-11738-1

Vol. 373. Oleg Okun, Giorgio Valentini, and Matteo Re (Eds.)
*Ensembles in Machine Learning Applications*, 2011
ISBN 978-3-642-22909-1

Vol. 374. Dimitri Plemenos and Georgios Miaoulis (Eds.)
*Intelligent Computer Graphics 2011*, 2011
ISBN 978-3-642-22906-0

Vol. 375. Marenglen Biba and Fatos Xhafa (Eds.)
*Learning Structure and Schemas from Documents*, 2011
ISBN 978-3-642-22912-1

Vol. 376. Toyohide Watanabe and Lakhmi C. Jain (Eds.)
*Innovations in Intelligent Machines - 2*, 2012
ISBN 978-3-642-23189-6

Vol. 377. Roger Lee (Ed.)
*Software Engineering Research, Management and Applications 2011*, 2011
ISBN 978-3-642-23201-5

Vol. 378. János Fodor, Ryszard Klempous, and Carmen Paz Suárez Araujo (Eds.)
*Recent Advances in Intelligent Engineering Systems*, 2011
ISBN 978-3-642-23228-2

Vol. 379. Ferrante Neri, Carlos Cotta, and Pablo Moscato (Eds.)
*Handbook of Memetic Algorithms*, 2012
ISBN 978-3-642-23246-6

Vol. 380. Anthony Brabazon, Michael O'Neill, and Dietmar Maringer (Eds.)
*Natural Computing in Computational Finance*, 2011
ISBN 978-3-642-23335-7

Vol. 381. Radosław Katarzyniak, Tzu-Fu Chiu, Chao-Fu Hong, and Ngoc Thanh Nguyen (Eds.)
*Semantic Methods for Knowledge Management and Communication*, 2011
ISBN 978-3-642-23417-0

Vol. 382. F.M.T. Brazier, Kees Nieuwenhuis, Gregor Pavlin, Martijn Warnier, and Costin Badica (Eds.)
*Intelligent Distributed Computing V*, 2011
ISBN 978-3-642-24012-6

Vol. 383. Takayuki Ito, Minjie Zhang, Valentin Robu, Shaheen Fatima, and Tokuro Matsuo(Eds.)
*New Trends in Agent-based Complex Automated Negotiations*, 2011
ISBN 978-3-642-24695-1

Vol. 384. Daphna Weinshall, Jörn Anemüller, and Luc van Gool (Eds.)
*Detection and Identification of Rare Audiovisual Cues*, 2012
ISBN 978-3-642-24033-1

Vol. 385. Alex Graves
*Supervised Sequence Labelling with Recurrent Neural Networks*, 2012
ISBN 978-3-642-24796-5

Alex Graves

# Supervised Sequence Labelling with Recurrent Neural Networks

Springer

*Author*
Dr. Alex Graves
University of Toronto
Department of Computer Science
Toronto, Ontario
Canada

ISSN 1860-949X e-ISSN 1860-9503
ISBN 978-3-642-24796-5 e-ISBN 978-3-642-24797-2
DOI 10.1007/978-3-642-24797-2
Springer Heidelberg New York Dordrecht London

Library of Congress Control Number: 2011940011



Printed on acid-free paper

Springer is part of Springer Science+Business Media (www.springer.com)

# Contents

# List of Tables

# List of Figures

# List of Algorithms

# Chapter 1
# Introduction

In machine learning, the term *sequence labelling* encompasses all tasks where sequences of data are transcribed with sequences of discrete labels. Well-known examples include speech and handwriting recognition, protein secondary structure prediction and part-of-speech tagging. *Supervised* sequence labelling refers specifically to those cases where a set of hand-transcribed sequences is provided for algorithm training. What distinguishes such problems from the traditional framework of *supervised pattern classification* is that the individual data points cannot be assumed to be independent. Instead, both the inputs and the labels form strongly correlated sequences. In speech recognition for example, the input (a speech signal) is produced by the continuous motion of the vocal tract, while the labels (a sequence of words) are mutually constrained by the laws of syntax and grammar. A further complication is that in many cases the alignment between inputs and labels is unknown. This requires the use of algorithms able to determine the location as well as the identity of the output labels.

*Recurrent neural networks* (RNNs) are a class of artificial neural network architecture that—inspired by the cyclical connectivity of neurons in the brain—uses iterative function loops to store information. RNNs have several properties that make them an attractive choice for sequence labelling: they are flexible in their use of context information (because they can learn what to store and what to ignore); they accept many different types and representations of data; and they can recognise sequential patterns in the presence of sequential distortions. However they also have several drawbacks that have limited their application to real-world sequence labelling problems.

Perhaps the most serious flaw of standard RNNs is that it is very difficult to get them to store information for long periods of time (Hochreiter et al., 2001b). This limits the range of context they can access, which is of critical importance to sequence labelling. *Long Short-Term Memory* (LSTM; Hochreiter and Schmidhuber, 1997) is a redesign of the RNN architecture around special 'memory cell' units. In various synthetic tasks, LSTM has been shown capable of storing and accessing information over very long timespans (Gers et al., 2002; Gers and Schmidhuber, 2001). It has also proved advantageous in real-world domains such as speech

A. Graves: Supervised Sequence Labell. with Recur. Neur. Networks, SCI 385, pp. 1–3.
springerlink.com 

processing (Graves and Schmidhuber, 2005b) and bioinformatics (Hochreiter et al., 2007). LSTM is therefore the architecture of choice throughout the book.

Another issue with the standard RNN architecture is that it can only access contextual information in one direction (typically the past, if the sequence is temporal). This makes perfect sense for time-series prediction, but for sequence labelling it is usually advantageous to exploit the context on both sides of the labels. *Bidirectional RNNs* (Schuster and Paliwal, 1997) scan the data forwards and backwards with two separate recurrent layers, thereby removing the asymmetry between input directions and providing access to all surrounding context. *Bidirectional LSTM* (Graves and Schmidhuber, 2005b) combines the benefits of long-range memory and bidirectional processing.

For tasks such as speech recognition, where the alignment between the inputs and the labels is unknown, RNNs have so far been limited to an auxiliary role. The problem is that the standard training methods require a separate target for every input, which is usually not available. The traditional solution—the so-called *hybrid approach*—is to use hidden Markov models to generate targets for the RNN, then invert the RNN outputs to provide observation probabilities (Bourlard and Morgan, 1994). However the hybrid approach does not exploit the full potential of RNNs for sequence processing, and it also leads to an awkward combination of discriminative and generative training. The *connectionist temporal classification* (CTC) output layer (Graves et al., 2006) removes the need for hidden Markov models by directly training RNNs to label sequences with unknown alignments, using a single discriminative loss function. CTC can also be combined with probabilistic language models for word-level speech and handwriting recognition.

Recurrent neural networks were designed for one-dimensional sequences. However some of their properties, such as robustness to warping and flexible use of context, are also desirable in multidimensional domains like image and video processing. *Multidimensional RNNs*, a special case of *directed acyclic graph RNNs* (Baldi and Pollastri, 2003), generalise to multidimensional data by replacing the one-dimensional chain of network updates with an n-dimensional grid. *Multidimensional LSTM* (Graves et al., 2007) brings the improved memory of LSTM to multidimensional networks.

Even with the LSTM architecture, RNNs tend to struggle with very long data sequences. As well as placing increased demands on the network's memory, such sequences can be be prohibitively time-consuming to process. The problem is especially acute for multidimensional data such as images or videos, where the volume of input information can be enormous. *Hierarchical subsampling RNNs* (Graves and Schmidhuber, 2009) contain a stack of recurrent network layers with progressively lower spatiotemporal resolution. As long as the reduction in resolution is large enough, and the layers at the bottom of the hierarchy are small enough, this approach can be made computationally efficient for almost any size of sequence. Furthermore, because the effective distance between the inputs decreases as the information moves up the hierarchy, the network's memory requirements are reduced.

The combination of multidimensional LSTM, CTC output layers and hierarchical subsampling leads to a general-purpose sequence labelling system entirely constructed out of recurrent neural networks. The system is flexible, and can be applied with minimal adaptation to a wide range of data and tasks. It is also powerful, as this book will demonstrate with state-of-the-art results in speech and handwriting recognition.

## 1.1 Structure of the Book

The chapters are roughly grouped into three parts: background material is presented in Chapters 2–4, Chapters 5 and 6 are primarily experimental, and new methods are introduced in Chapters 7–9.

Chapter 2 briefly reviews supervised learning in general, and pattern classification in particular. It also provides a formal definition of sequence labelling, and discusses three classes of sequence labelling task that arise under different relationships between the input and label sequences. Chapter 3 provides background material for feedforward and recurrent neural networks, with emphasis on their application to labelling and classification tasks. It also introduces the *sequential Jacobian* as a tool for analysing the use of context by RNNs.

Chapter 4 describes the LSTM architecture and introduces bidirectional LSTM (BLSTM). Chapter 5 contains an experimental comparison of BLSTM to other neural network architectures applied to framewise phoneme classification. Chapter 6 investigates the use of LSTM in hidden Markov model-neural network hybrids. Chapter 7 introduces connectionist temporal classification, Chapter 8 covers multidimensional networks, and hierarchical subsampling networks are described in Chapter 9.

# Chapter 2
# Supervised Sequence Labelling

This chapter provides the background material and literature review for supervised sequence labelling. Section 2.1 briefly reviews supervised learning in general. Section 2.2 covers the classical, non-sequential framework of supervised pattern classification. Section 2.3 defines supervised sequence labelling, and describes the different classes of sequence labelling task that arise under different assumptions about the label sequences.

## 2.1 Supervised Learning

Machine learning problems where a set of input-target pairs is provided for training are referred to as *supervised learning* tasks. This is distinct from *reinforcement learning*, where only scalar reward values are provided for training, and *unsupervised learning*, where no training signal exists at all, and the algorithm attempts to uncover the structure of the data by inspection alone. We will not consider either reinforcement learning or unsupervised learning in this book.

A supervised learning task consists of a *training set* $S$ of input-target pairs $(x, z)$, where $x$ is an element of the input space $\mathcal{X}$ and $z$ is an element of the target space $\mathcal{Z}$, along with a disjoint *test set* $S'$. We will sometimes refer to the elements of $S$ as *training examples*. Both $S$ and $S'$ are assumed to have been drawn independently from the same input-target distribution $\mathcal{D}_{\mathcal{X}\times\mathcal{Z}}$. In some cases an extra *validation set* is drawn from the training set to validate the performance of the learning algorithm during training; in particular validation sets are frequently used to determine when training should stop, in order to prevent overfitting. The goal is to use the training set to minimise some task-specific error measure $E$ defined on the test set. For example, in a regression task, the usual error measure is the *sum-of-squares*, or squared Euclidean distance between the algorithm outputs and the test-set targets. For parametric algorithms (such as neural networks) the usual approach to error minimisation is to incrementally adjust the algorithm parameters to optimise a *loss function* on the training set, which is as closely related as possible to $E$. The transfer of learning from the training set to

A. Graves: Supervised Sequence Labell. with Recur. Neur. Networks, SCI 385, pp. 5–13.
springerlink.com 

the test set is known as *generalisation*, and will be discussed further in later chapters.

The nature and degree of supervision provided by the targets varies greatly between supervised learning tasks. For example, training a supervised learner to correctly label every pixel corresponding to an aeroplane in an image requires a much more informative target than simply training it recognise whether or not an aeroplane is present. To distinguish these extremes, people sometimes refer to *weakly* and *strongly* labelled data.

## 2.2 Pattern Classification

Pattern classification, also known as pattern recognition, is one of the most extensively studied areas of machine learning (Bishop, 2006; Duda et al., 2000), and certain pattern classifiers, such as multilayer perceptrons (Rumelhart et al., 1986; Bishop, 1995) and support vector machines (Vapnik, 1995) have become familiar to the scientific community at large.

Although pattern classification deals with non-sequential data, much of the practical and theoretical framework underlying it carries over to the sequential case. It is therefore instructive to briefly review this framework before we turn to sequence labelling.

The input space $\mathcal{X}$ for supervised pattern classification tasks is typically $\mathbf{R}^M$; that is, the set of all real-valued vectors of some fixed length $M$. The target spaces $\mathcal{Z}$ is a discrete set of $K$ classes. A pattern classifier $h : \mathcal{X} \mapsto \mathcal{Z}$ is therefore a function mapping from vectors to labels. If all misclassifications are equally bad, the usual error measure for $h$ is the *classification error rate* $E^{class}(h, S')$ on the test set $S'$

$$E^{class}(h, S') = \frac{1}{|S'|} \sum_{(x,z) \in S'} \begin{cases} 0 \text{ if } h(x) = z \\ 1 \text{ otherwise} \end{cases} \tag{2.1}$$

### 2.2.1 *Probabilistic Classification*

Classifiers that directly output class labels, of which support vector machines are a well known example, are sometimes referred to as *discriminant functions*. An alternative approach is *probabilistic classification*, where the conditional probabilities $p(C_k|x)$ of the $K$ classes given the input pattern $x$ are first determined, and the most probable is then chosen as the classifier output $h(x)$:

$$h(x) = \arg\max_k p(C_k|x) \tag{2.2}$$

One advantage of the probabilistic approach is that the relative magnitude of the probabilities can be used to determine the degree of confidence the classifier has in its outputs. Another is that it allows the classifier to be combined with other probabilistic algorithms in a consistent way.

### 2.2.2 *Training Probabilistic Classifiers*

If a probabilistic classifier $h_w$ yields a conditional distribution $p(C_k|x, w)$ over the class labels $C_k$ given input $x$ and parameters $w$, we can take a product over the independent and identically distributed (i.i.d.) input-target pairs in the training set $S$ to get

$$p(S|w) = \prod_{(x,z)\in S} p(z|x, w) \tag{2.3}$$

which can be inverted with Bayes' rule to obtain

$$p(w|S) = \frac{p(S|w)p(w)}{p(S)} \tag{2.4}$$

In theory, the posterior distribution over classes for some new input $x$ can then be found by integrating over all possible values of $w$:

$$p(C_k|x, S) = \int_w p(C_k|x, w)p(w|S)dw \tag{2.5}$$

In practice $w$ is usually very high dimensional and the above integral, referred to as the *predictive distribution* of the classifier, is intractable. A common approximation, known as the maximum a priori (MAP) approximation, is to find the single parameter vector $w_{MAP}$ that maximises $p(w|S)$ and use this to make predictions:

$$p(C_k|x, S) \approx p(C_k|x, w_{MAP}) \tag{2.6}$$

Since $p(S)$ is independent of $w$, Eqn. (2.4) tells us that

$$w_{MAP} = \arg\max_w p(S|w)p(w) \tag{2.7}$$

The parameter prior $p(w)$ is usually referred to as a *regularisation term*. Its effect is to weight the classifier towards those parameter values which are deemed *a priori* more probable. In accordance with Occam's razor, we usually assume that more complex parameters (where 'complex' is typically interpreted as 'requiring more information to accurately describe') are inherently less probable. For this reason $p(w)$ is sometimes referred to as an *Occam factor* or *complexity penalty*. In the particular case of a Gaussian parameter prior, where $p(w) \propto |w|^2$, the $p(w)$ term is referred to as *weight decay*. If, on the other hand, we assume a uniform prior over parameters, we can remove the $p(w)$ term from (2.7) to obtain the maximum likelihood (ML) parameter vector $w_{ML}$

$$w_{ML} = \arg\max_w p(S|w) = \arg\max_w \prod_{(x,z)\in S} p(z|x, w) \tag{2.8}$$

From now on we will drop the explicit dependence of the classifier outputs on $w$, with the understanding that $p(z|x)$ is the probability of $x$ being correctly classified by $h_w$.

#### 2.2.2.1 Maximum-Likelihood Loss Functions

The standard procedure for finding $w_{ML}$ is to minimise a maximum-likelihood loss function $\mathcal{L}(S)$ defined as the negative logarithm of the probability assigned to $S$ by the classifier

$$\mathcal{L}(S) = -\ln \prod_{(x,z)\in S} p(z|x) = -\sum_{(x,z)\in S} \ln p(z|x) \tag{2.9}$$

where ln is the *natural logarithm* (the logarithm to base $e$). Note that, since the logarithm is monotonically increasing, minimising $-\ln p(S)$ is equivalent to maximising $p(S)$.

Observing that each example training example $(x, z) \in S$ contributes to a single term in the above sum, we define the *example loss* $\mathcal{L}(x, z)$ as

$$\mathcal{L}(x, z) = -\ln p(z|x) \tag{2.10}$$

and note that

$$\mathcal{L}(S) = \sum_{(x,z)\in S} \mathcal{L}(x, z) \tag{2.11}$$

$$\frac{\partial \mathcal{L}(S)}{\partial w} = \sum_{(x,z)\in S} \frac{\partial \mathcal{L}(x, z)}{\partial w} \tag{2.12}$$

It therefore suffices to derive $\mathcal{L}(x, z)$ and $\frac{\partial \mathcal{L}(x,z)}{\partial w}$ to completely define a maximum-likelihood loss function and its derivatives with respect to the network weights.

When the precise form of the loss function is not important, we will refer to it simply as $\mathcal{L}$.

### *2.2.3 Generative and Discriminative Methods*

Algorithms that directly calculate the class probabilities $p(C_k|x)$ (also known as the posterior class probabilities) are referred to as *discriminative*. In some cases however, it is preferable to first calculate the class conditional densities $p(x|C_k)$ and then use Bayes' rule, together with the prior class probabilities $p(C_k)$ to find the posterior values

$$p(C_k|x) = \frac{p(x|C_k)p(C_k)}{p(x)} \tag{2.13}$$

where

$$p(x) = \sum_k p(x|C_k)p(C_k) \tag{2.14}$$

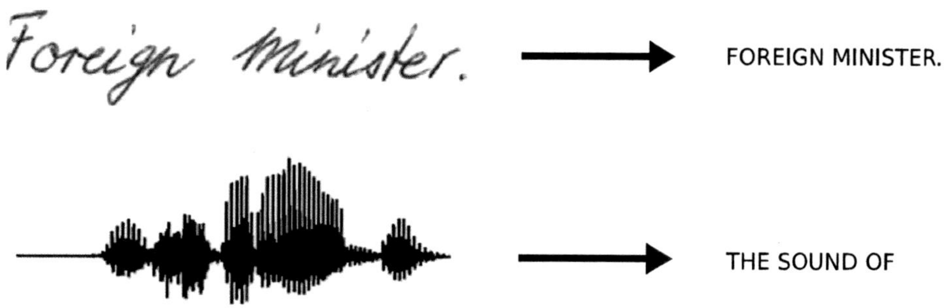


**Fig. 2.1 Sequence labelling.** The algorithm receives a sequence of input data, and outputs a sequence of discrete labels.

Algorithms following this approach are referred to as *generative*, because the prior $p(x)$ can be used to generate artificial input data. One advantage of the generative approach is that each class can be trained independently of the others, whereas discriminative methods have to be retrained every time a new class is added. However, discriminative methods typically give better results for classification tasks, because they concentrate all their modelling effort on finding the correct class boundaries.

This book focuses on discriminative sequence labelling. However, we will frequently refer to the well-known generative method *hidden Markov models* (Rabiner, 1989; Bengio, 1999).

## 2.3 Sequence Labelling

The goal of sequence labelling is to assign sequences of labels, drawn from a fixed alphabet, to sequences of input data. For example, one might wish to transcribe a sequence of acoustic features with spoken words (speech recognition), or a sequence of video frames with hand gestures (gesture recognition). Although such tasks commonly arise when analysing time series, they are also found in domains with non-temporal sequences, such as protein secondary structure prediction.

For some problems the precise alignment of the labels with the input data must also be determined by the learning algorithm. In this book however, we limit our attention to tasks where the alignment is either predetermined, by some manual or automatic preprocessing, or it is unimportant, in the sense that we require only the final *sequence* of labels, and not the times at which they occur.

If the sequences are assumed to be independent and identically distributed, we recover the basic framework of pattern classification, only with sequences in place of patterns (of course the data-points within each sequence are *not* assumed to be independent). In practice this assumption may not be

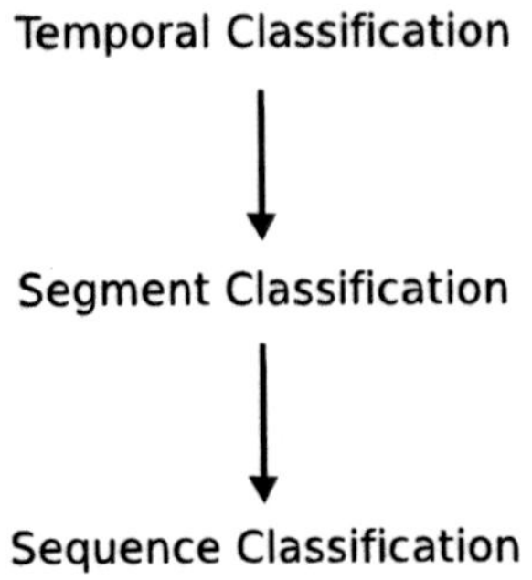


**Fig. 2.2 Three classes of sequence labelling task.** Sequence classification, where each input sequence is assigned a single class, is a special case of segment classification, where each of a predefined set of input segments is given a label. Segment classification is a special case of temporal classification, where any alignment between input and label sequences is allowed. Temporal classification data can be *weakly labelled* with nothing but the target sequences, while segment classification data must be *strongly labelled* with both targets and input-target alignments.

entirely justified (for example, the sequences may represent turns in a spoken dialogue, or lines of text in a handwritten form); however it is usually not too damaging as long as the sequence boundaries are sensibly chosen. We further assume that each target sequence is at most as long as the corresponding input sequence. With these restrictions in mind we can formalise the task of sequence labelling as follows:

Let $S$ be a set of training examples drawn independently from a fixed distribution $\mathcal{D}_{\mathcal{X}\times\mathcal{Z}}$. The input space $\mathcal{X} = (\mathbb{R}^M)^*$ is the set of all sequences of size $M$ real-valued vectors. The target space $\mathcal{Z} = L^*$ is the set of all sequences over the (finite) alphabet $L$ of labels. We refer to elements of $L^*$ as *label sequences* or *labellings*. Each element of $S$ is a pair of sequences $(\mathbf{x}, \mathbf{z})$ (From now on a bold typeface will be used to denote sequences). The target sequence $\mathbf{z} = (z_1, z_2, ..., z_U)$ is at most as long as the input sequence $\mathbf{x} = (x_1, x_2, ..., x_T)$, i.e. $|\mathbf{z}| = U \leq |\mathbf{x}| = T$. Regardless of whether the data is a time series, the distinct points in the *input* sequence are referred to as *timesteps*. The task is to use $S$ to train a sequence labelling algorithm $h : \mathcal{X} \mapsto \mathcal{Z}$ to label the sequences in a test set $S' \subset \mathcal{D}_{\mathcal{X}\times\mathcal{Z}}$, disjoint from $S$, as accurately as possible.

In some cases we can apply additional constraints to the label sequences. These may affect both the choice of sequence labelling algorithm and the error measures used to assess performance. The following sections describe three classes of sequence labelling task, corresponding to progressively looser assumptions about the relationship between the input and label sequences, and discuss algorithms and error measures suitable for each. The relationship between the classes is outlined in Figure 2.2.

### 2.3.1 Sequence Classification

The most restrictive case is where the label sequences are constrained to be length one. This is referred to as *sequence classification*, since each input sequence is assigned to a single class. Examples of sequence classification task include the identification of a single spoken work and the recognition of an individual handwritten letter. A key feature of such tasks is that the entire sequence can be processed before the classification is made.

If the input sequences are of fixed length, or can be easily padded to a fixed length, they can be collapsed into a single input vector and any of the standard pattern classification algorithms mentioned in Section 2.2 can be applied. A prominent testbed for fixed-length sequence classification is the MNIST isolated digits dataset (LeCun et al., 1998a). Numerous pattern classification algorithms have been applied to MNIST, including convolutional neural networks (LeCun et al., 1998a; Simard et al., 2003) and support vector machines (LeCun et al., 1998a; Decoste and Schölkopf, 2002).

However, even if the input length is fixed, algorithm that are inherently sequential may be beneficial, since they are better able to adapt to translations and distortions in the input data. This is the rationale behind the application of multidimensional recurrent neural networks to MNIST in Chapter 8.

As with pattern classification the obvious error measure is the percentage of misclassifications, referred to as the *sequence error rate* $E^{seq}$ in this context:

$$E^{seq}(h, S') = \frac{100}{|S'|} \sum_{(\mathbf{x},\mathbf{z})\in S'} \begin{cases} 0 \text{ if } h(\mathbf{x}) = \mathbf{z} \\ 1 \text{ otherwise} \end{cases} \tag{2.15}$$

where $|S'|$ is the number of elements in $S'$.

### 2.3.2 Segment Classification

*Segment classification* refers to those tasks where the target sequences consist of multiple labels, but the locations of the labels—that is, the positions of the input segments to which the labels apply—are known in advance. Segment classification is common in domains such as natural language processing and bioinformatics, where the inputs are discrete and can be trivially segmented. It can also occur in domains where segmentation is difficult, such as audio or image processing; however this typically requires hand-segmented training data, which is difficult to obtain.

A crucial element of segment classification, missing from sequence classification, is the use of *context* information from either side of the segments to be classified. The effective use of context is vital to the success of segment classification algorithms, as illustrated in Figure 2.3. This presents a problem for standard pattern classification algorithms, which are designed to process only one input at a time. A simple solution is to collect the data on either side of the segments into *time-windows*, and use the windows as input

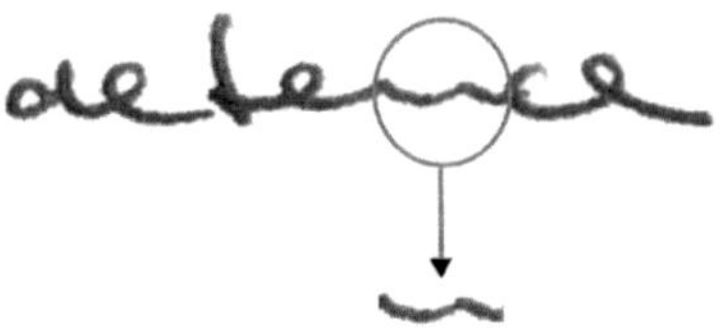

**Fig. 2.3 Importance of context in segment classification.** The word 'defence' is clearly legible. However the letter 'n' in isolation is ambiguous.

patterns. However as well as the aforementioned issue of shifted or distorted data, the time-window approach suffers from the fact that the range of useful context (and therefore the required time-window size) is generally unknown, and may vary from segment to segment. Consequently the case for sequential algorithms is stronger here than in sequence classification.

The obvious error measure for segment classification is the *segment error rate* $E^{seg}$, which simply counts the percentage of misclassified segments.

$$E^{seg}(h, S') = \frac{1}{Z} \sum_{(\mathbf{x},\mathbf{z}) \in S'} HD(h(\mathbf{x}), \mathbf{z}) \tag{2.16}$$

Where

$$Z = \sum_{(\mathbf{x},\mathbf{z}) \in S'} |z| \tag{2.17}$$

and $HD(\mathbf{p}, \mathbf{q})$ is the hamming distance between two equal length sequences $\mathbf{p}$ and $\mathbf{q}$ (i.e. the number of places in which they differ).

In speech recognition, the phonetic classification of each acoustic frame as a separate segment is often known as *framewise phoneme classification.* In this context the segment error rate is usually referred to as the *frame error rate.* Various neural network architectures are applied to framewise phoneme classification in Chapter 5. In image processing, the classification of each pixel, or block of pixels, as a separate segment is known as *image segmentation.* Multidimensional recurrent neural networks are applied to image segmentation in Chapter 8.

### *2.3.3 Temporal Classification*

In the most general case, nothing can be assumed about the label sequences except that their length is less than or equal to that of the input sequences. They may even be empty. We refer to this situation as *temporal classification* (Kadous, 2002).

The key distinction between temporal classification and segment classification is that the former requires an algorithm that can decide *where* in the input sequence the classifications should be made. This in turn requires an implicit or explicit model of the global structure of the sequence.

For temporal classification, the segment error rate is inapplicable, since the segment boundaries are unknown. Instead we measure the total number of substitutions, insertions and deletions that would be required to turn one sequence into the other, giving us the *label error rate* $E^{lab}$:

$$E^{lab}(h, S') = \frac{1}{Z} \sum_{(\mathbf{x},\mathbf{z}) \in S'} ED(h(\mathbf{x}), \mathbf{z}) \tag{2.18}$$

Where $ED(\mathbf{p}, \mathbf{q})$ is the *edit distance* between the two sequences $\mathbf{p}$ and $\mathbf{q}$ (i.e. the minimum number of insertions, substitutions and deletions required to change $\mathbf{p}$ into $\mathbf{q}$). $ED(\mathbf{p}, \mathbf{q})$ can be calculated in $O(|\mathbf{p}||\mathbf{q}|)$ time (Navarro, 2001). The label error rate is typically multiplied by 100 so that it can be interpreted as a percentage (a convention we will follow in this book); however, unlike the other error measures considered in this chapter, it is not a true percentage, and may give values higher than 100.

A family of similar error measures can be defined by introducing other types of edit operation, such as transpositions (caused by e.g. typing errors), or by weighting the relative importance of the operations. For the purposes of this book however, the label error rate is sufficient. We will usually refer to the label error rate according to the type of label in question, for example *phoneme error rate* or *word error rate*. For some temporal classification tasks a completely correct labelling is required and the degree of error is unimportant. In this case the sequence error rate (2.15) should be used to assess performance.

The use of hidden Markov model-recurrent neural network hybrids for temporal classification is investigated in Chapter 6, and a neural-network-only approach to temporal classification is introduced in Chapter 7.

# Chapter 3
# Neural Networks

This chapter provides an overview of artificial neural networks, with emphasis on their application to classification and labelling tasks. Section 3.1 reviews multilayer perceptrons and their application to pattern classification. Section 3.2 reviews recurrent neural networks and their application to sequence labelling. It also describes the sequential Jacobian, an analytical tool for studying the use of context information. Section 3.3 discusses various issues, such as generalisation and input data representation, that are essential to effective network training.

## 3.1 Multilayer Perceptrons

*Artificial neural networks* (ANNs) were originally developed as mathematical models of the information processing capabilities of biological brains (McCulloch and Pitts, 1988, Rosenblatt, 1963; Rumelhart et al., 1986). Although it is now clear that ANNs bear little resemblance to real biological neurons, they enjoy continuing popularity as pattern classifiers.

The basic structure of an ANN is a network of small processing units, or nodes, joined to each other by weighted connections. In terms of the original biological model, the nodes represent neurons, and the connection weights represent the strength of the synapses between the neurons. The network is activated by providing an input to some or all of the nodes, and this activation then spreads throughout the network along the weighted connections. The electrical activity of biological neurons typically follows a series of sharp 'spikes', and the activation of an ANN node was originally intended to model the average firing rate of these spikes.

Many varieties of ANNs have appeared over the years, with widely varying properties. One important distinction is between ANNs whose connections form cycles, and those whose connections are acyclic. ANNs with cycles are referred to as feedback, recursive, or recurrent, neural networks, and are dealt with in Section 3.2. ANNs without cycles are referred to as feedforward neural networks (FNNs). Well known examples of FNNs include *perceptrons* (Rosenblatt, 1958), *radial basis function networks* (Broomhead and Lowe,

A. Graves: Supervised Sequence Labell. with Recur. Neur. Networks, SCI 385, pp. 15–35.
springerlink.com

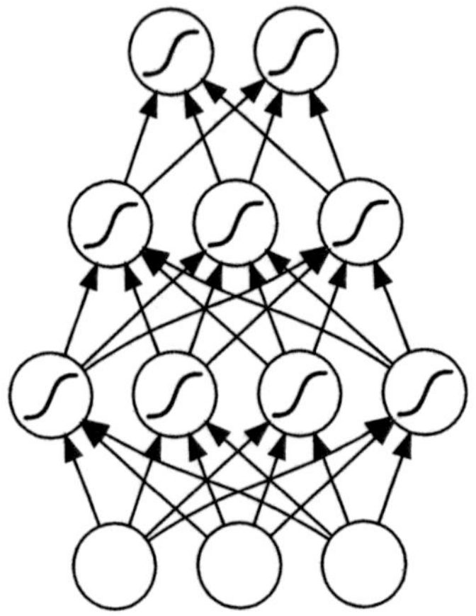


**Fig. 3.1 A multilayer perceptron.** The S-shaped curves in the hidden and output layers indicate the application of 'sigmoidal' nonlinear activation functions.

1988), *Kohonen maps* (Kohonen, 1989) and *Hopfield nets* (Hopfield, 1982). The most widely used form of FNN, and the one we focus on in this section, is the *multilayer perceptron* (MLP; Rumelhart et al., 1986; Werbos, 1988; Bishop, 1995).

As illustrated in Figure 3.1, the units in a multilayer perceptron are arranged in layers, with connections feeding forward from one layer to the next. Input patterns are presented to the *input layer*, then propagated through the *hidden layers* to the *output layer*. This process is known as the *forward pass* of the network.

Since the output of an MLP depends only on the current input, and not on any past or future inputs, MLPs are more suitable for pattern classification than for sequence labelling. We will discuss this point further in Section 3.2.

An MLP with a particular set of weight values defines a function from input to output vectors. By altering the weights, a single MLP is capable of instantiating many different functions. Indeed it has been proven (Hornik et al., 1989) that an MLP with a single hidden layer containing a sufficient number of nonlinear units can approximate *any* continuous function on a compact input domain to arbitrary precision. For this reason MLPs are said to be *universal function approximators.*

### *3.1.1 Forward Pass*

Consider an MLP with $I$ input units, activated by input vector $x$ (hence $|x| = I$). Each unit in the first hidden layer calculates a weighted sum of the input units. For hidden unit $h$, we refer to this sum as the *network input* to unit $h$, and denote it $a_h$. The *activation function* $\theta_h$ is then applied, yielding the final activation $b_h$ of the unit. Denoting the weight from unit $i$ to unit $j$ as $w_{ij}$, we have

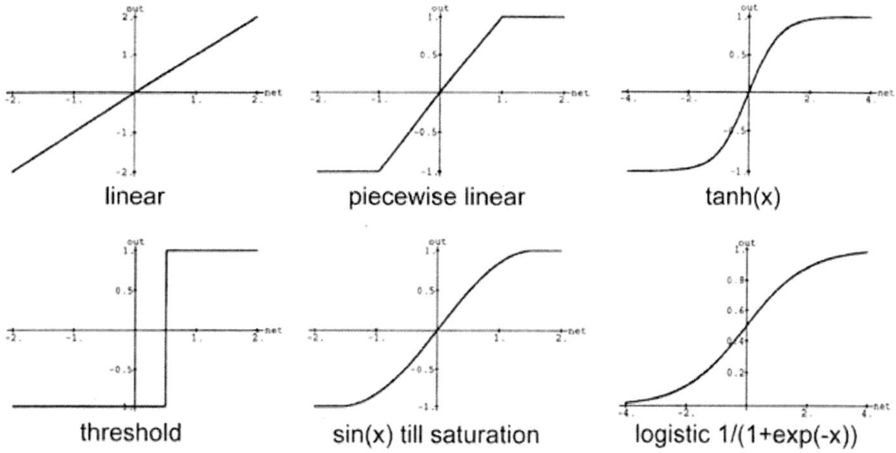


**Fig. 3.2 Neural network activation functions.** Note the characteristic 'sigmoid' or S-shape.

$$a_h = \sum_{i=1}^{I} w_{ih} x_i \tag{3.1}$$

$$b_h = \theta_h(a_h) \tag{3.2}$$

Several neural network activation functions are plotted in Figure 3.2. The most common choices are the hyperbolic tangent

$$tanh(x) = \frac{e^{2x} - 1}{e^{2x} + 1}, \tag{3.3}$$

and the logistic sigmoid

$$\sigma(x) = \frac{1}{1 + e^{-x}} \tag{3.4}$$

The two functions are related by the following linear transform:

$$tanh(x) = 2\sigma(2x) - 1 \tag{3.5}$$

This means that any function computed by a neural network with a hidden layer of tanh units can be computed by another network with logistic sigmoid units and vice-versa. They are therefore largely equivalent as activation functions. However one reason to distinguish between them is that their output ranges are different; in particular if an output between 0 and 1 is required (for example, if the output represents a probability) then the logistic sigmoid should be used.

An important feature of both tanh and the logistic sigmoid is their nonlinearity. Nonlinear neural networks are more powerful than linear ones since they can, for example, find nonlinear classification boundaries and model nonlinear equations. Moreover, any combination of linear operators is itself a linear operator, which means that any MLP with multiple linear hidden layers is exactly equivalent to some other MLP with a single linear hidden layer. This contrasts with nonlinear networks, which can gain considerable power by using successive hidden layers to re-represent the input data (Hinton et al., 2006; Bengio and LeCun, 2007).

Another key property is that both functions are differentiable, which allows the network to be trained with gradient descent. Their first derivatives are

$$\frac{\partial tanh(x)}{\partial x} = 1 - tanh(x)^2 \tag{3.6}$$

$$\frac{\partial \sigma(x)}{\partial x} = \sigma(x)(1 - \sigma(x)) \tag{3.7}$$

Because of the way they reduce an infinite input domain to a finite output range, neural network activation functions are sometimes referred to as *squashing functions*.

Having calculated the activations of the units in the first hidden layer, the process of summation and activation is then repeated for the rest of the hidden layers in turn, e.g. for unit $h$ in the $l^{th}$ hidden layer $H_l$

$$a_h = \sum_{h' \in H_{l-1}} w_{h'h} b_{h'} \tag{3.8}$$

$$b_h = \theta_h(a_h) \tag{3.9}$$

### 3.1.2 Output Layers

The output vector $y$ of an MLP is given by the activation of the units in the output layer. The network input $a_k$ to each output unit $k$ is calculated by summing over the units connected to it, exactly as for a hidden unit. Therefore

$$a_k = \sum_{h \in H_L} w_{hk} b_h \tag{3.10}$$

for a network with $L$ hidden layers.

Both the number of units in the output layer and the choice of output activation function depend on the task the network is applied to. For binary classification tasks, the standard configuration is a single unit with a logistic sigmoid activation (Eqn. (3.4)). Since the range of the logistic sigmoid is the open interval $(0, 1)$, the activation of the output unit can be interpreted as the probability that the input vector belongs to the first class (and conversely, one minus the activation gives the probability that it belongs to the second class)

$$\begin{aligned} p(C_1|x) &= y = \sigma(a) \\ p(C_2|x) &= 1 - y \end{aligned} \tag{3.11}$$

The use of the logistic sigmoid as a binary probability estimator is sometimes referred as *logistic regression*, or a *logit* model. If we use a coding scheme for the target vector $z$ where $z = 1$ if the correct class is $C_1$ and $z = 0$ if the correct class is $C_2$, we can combine the above expressions to write

$$p(z|x) = y^z(1 - y)^{1-z} \tag{3.12}$$

For classification problems with $K > 2$ classes, the convention is to have $K$ output units, and normalise the output activations with the *softmax* function (Bridle, 1990) to obtain the class probabilities:

$$p(C_k|x) = y_k = \frac{e^{a_k}}{\sum_{k'=1}^{K} e^{a_{k'}}} \tag{3.13}$$

which is also known as a *multinomial logit* model. A *1-of-K* coding scheme represent the target class $z$ as a binary vector with all elements equal to zero except for element $k$, corresponding to the correct class $C_k$, which equals one. For example, if $K = 5$ and the correct class is $C_2$, $z$ is represented by $(0, 1, 0, 0, 0)$. Using this scheme we obtain the following convenient form for the target probabilities:

$$p(z|x) = \prod_{k=1}^{K} y_k^{z_k} \tag{3.14}$$

Given the above definitions, the use of MLPs for pattern classification is straightforward. Simply feed in an input vector, activate the network, and choose the class label corresponding to the most active output unit.

### *3.1.3 Loss Functions*

The derivation of loss functions for MLP training follows the steps outlined in Section 2.2.2. Although attempts have been made to approximate the full predictive distribution of Eqn. (2.5) for neural networks (MacKay, 1995; Neal, 1996), we will here focus on loss functions derived using maximum likelihood. For binary classification, substituting (3.12) into the maximum-likelihood example loss $\mathcal{L}(x, z) = -\ln p(z|x)$ described in Section 2.2.2.1, we have

$$\mathcal{L}(x, z) = (z - 1)\ln(1 - y) - z \ln y \tag{3.15}$$

Similarly, for problems with multiple classes, substituting (3.14) into (2.10) gives

$$\mathcal{L}(x, z) = -\sum_{k=1}^{K} z_k \ln y_k \tag{3.16}$$

See (Bishop, 1995, chap. 6) for more information on these and other MLP loss functions.

### 3.1.4 Backward Pass

Since MLPs are, by construction, differentiable operators, they can be trained to minimise any differentiable loss function using *gradient descent*. The basic idea of gradient descent is to find the derivative of the loss function with respect to each of the network weights, then adjust the weights in the direction of the negative slope. Gradient descent methods for training neural networks are discussed in more detail in Section 3.3.1.

To efficiently calculate the gradient, we use a technique known as *backpropagation* (Rumelhart et al., 1986; Williams and Zipser, 1995; Werbos, 1988). This is often referred to as the *backward pass* of the network.

Backpropagation is simply a repeated application of chain rule for partial derivatives. The first step is to calculate the derivatives of the loss function with respect to the output units. For a binary classification network, differentiating the loss function defined in (3.15) with respect to the network outputs gives

$$\frac{\partial \mathcal{L}(x,z)}{\partial y} = \frac{y-z}{y(1-y)} \tag{3.17}$$

The chain rule informs us that

$$\frac{\partial \mathcal{L}(x,z)}{\partial a} = \frac{\partial \mathcal{L}(x,z)}{\partial y}\frac{\partial y}{\partial a} \tag{3.18}$$

and we can then substitute (3.7), (3.11) and (3.17) into (3.18) to get

$$\frac{\partial \mathcal{L}(x,z)}{\partial a} = y - z \tag{3.19}$$

For a multiclass network, differentiating (3.16) gives

$$\frac{\partial \mathcal{L}(x,z)}{\partial y_k} = -\frac{z_k}{y_k} \tag{3.20}$$

Bearing in mind that the activation of each unit in a softmax layer depends on the network input to every unit in the layer, the chain rule gives us

$$\frac{\partial \mathcal{L}(x,z)}{\partial a_k} = \sum_{k'=1}^{K} \frac{\partial \mathcal{L}(x,z)}{\partial y_{k'}}\frac{\partial y_{k'}}{\partial a_k} \tag{3.21}$$

Differentiating (3.13) we obtain

$$\frac{\partial y_{k'}}{\partial a_k} = y_k \delta_{kk'} - y_k y_{k'} \tag{3.22}$$

and we can then substitute (3.22) and (3.20) into (3.21) to get

$$\frac{\partial \mathcal{L}(x,z)}{\partial a_k} = y_k - z_k \tag{3.23}$$

where we have the used the fact that $\sum_{k=1}^{K} z_k = 1$. Note the similarity to (3.19). The loss function is sometimes said to *match* the output layer activation function when the output derivative has this form (Schraudolph, 2002).

We now continue to apply the chain rule, working backwards through the hidden layers. At this point it is helpful to introduce the following notation:

$$\delta_j \stackrel{\text{def}}{=} \frac{\partial \mathcal{L}(x,z)}{\partial a_j} \tag{3.24}$$

where $j$ is any unit in the network. For the units in the last hidden layer, we have

$$\delta_h = \frac{\partial \mathcal{L}(x,z)}{\partial b_h}\frac{\partial b_h}{\partial a_h} = \frac{\partial b_h}{\partial a_h}\sum_{k=1}^{K}\frac{\partial \mathcal{L}(x,z)}{\partial a_k}\frac{\partial a_k}{\partial b_h} \tag{3.25}$$

where we have used the fact that $\mathcal{L}(x,z)$ depends only on each hidden unit $h$ through its influence on the output units. Differentiating (3.10) and (3.2) and substituting into (3.25) gives

$$\delta_h = \theta'(a_j)\sum_{k=1}^{K}\delta_k w_{hk} \tag{3.26}$$

The $\delta$ terms for each hidden layer $H_l$ before the last one can then be calculated recursively:

$$\delta_h = \theta'(a_h)\sum_{h' \in H_{l+1}}\delta_{h'} w_{hh'} \tag{3.27}$$

Once we have the $\delta$ terms for all the hidden units, we can use (3.1) to calculate the derivatives with respect to each of the network weights:

$$\frac{\partial \mathcal{L}(x,z)}{\partial w_{ij}} = \frac{\partial \mathcal{L}(x,z)}{\partial a_j}\frac{\partial a_j}{\partial w_{ij}} = \delta_j b_i \tag{3.28}$$

#### 3.1.4.1 Numerical Gradient

When implementing backpropagation, it is strongly recommended to check the weight derivatives numerically. This can be done by adding positive and negative perturbations to each weight and calculating the changes in the loss function:

$$\frac{\partial \mathcal{L}}{\partial w_{ij}} = \frac{\mathcal{L}(w_{ij}+\epsilon) - \mathcal{L}(w_{ij}-\epsilon)}{2\epsilon} + \mathcal{O}(\epsilon^2) \tag{3.29}$$

This technique is known as *symmetrical finite differences*. Note that setting $\epsilon$ too small leads to numerical underflows and decreased accuracy. The optimal value therefore depends on the floating point accuracy of a given implementation. For the systems we used, $\epsilon = 10^{-5}$ generally gave best results.

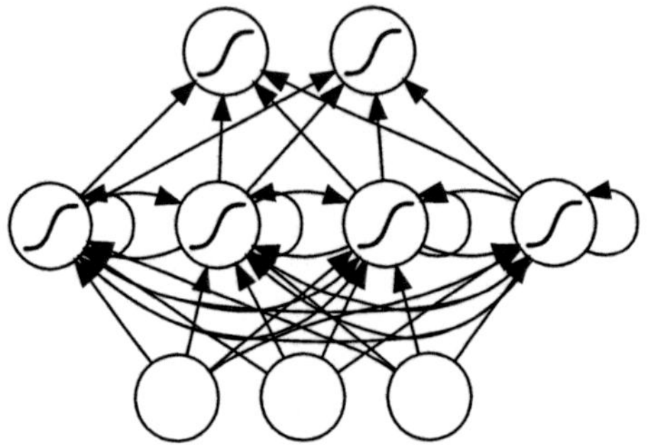


**Fig. 3.3 A recurrent neural network**

Note that for a network with $W$ weights, calculating the full gradient using (3.29) requires $O(W^2)$ time, whereas backpropagation only requires $O(W)$ time. Numerical differentiation is therefore impractical for network training. Furthermore, it is recommended to always the choose the smallest possible exemplar of the network architecture whose gradient you wish to check (for example, an RNN with a single hidden unit).

## 3.2 Recurrent Neural Networks

In the previous section we considered feedforward neural networks whose connections did not form cycles. If we relax this condition, and allow cyclical connections as well, we obtain *recurrent neural networks* (RNNs). As with feedforward networks, many varieties of RNN have been proposed, such as Elman networks (Elman, 1990), Jordan networks (Jordan, 1990), time delay neural networks (Lang et al., 1990) and echo state networks (Jaeger, 2001). In this chapter, we focus on a simple RNN containing a single, self connected hidden layer, as shown in Figure 3.3.

While the difference between a multilayer perceptron and an RNN may seem trivial, the implications for sequence learning are far-reaching. An MLP can only map from input to output vectors, whereas an RNN can in principle map from the entire *history* of previous inputs to each output. Indeed, the equivalent result to the universal approximation theory for MLPs is that an RNN with a sufficient number of hidden units can approximate any measurable sequence-to-sequence mapping to arbitrary accuracy (Hammer, 2000). The key point is that the recurrent connections allow a 'memory' of previous inputs to persist in the network's internal state, and thereby influence the network output.

### *3.2.1 Forward Pass*

The forward pass of an RNN is the same as that of a multilayer perceptron with a single hidden layer, except that activations arrive at the hidden layer from both the current external input and the hidden layer activations from the previous timestep. Consider a length $T$ input sequence $\mathbf{x}$ presented to an RNN with $I$ input units, $H$ hidden units, and $K$ output units. Let $x_i^t$ be the value of input $i$ at time $t$, and let $a_j^t$ and $b_j^t$ be respectively the network input to unit $j$ at time $t$ and the activation of unit $j$ at time $t$. For the hidden units we have

$$a_h^t = \sum_{i=1}^{I} w_{ih} x_i^t + \sum_{h'=1}^{H} w_{h'h} b_{h'}^{t-1} \tag{3.30}$$

Nonlinear, differentiable activation functions are then applied exactly as for an MLP

$$b_h^t = \theta_h(a_h^t) \tag{3.31}$$

The complete sequence of hidden activations can be calculated by starting at $t = 1$ and recursively applying (3.30) and (3.31), incrementing $t$ at each step. Note that this requires initial values $b_i^0$ to be chosen for the hidden units, corresponding to the network's state before it receives any information from the data sequence. In this book, the initial values are always set to zero. However, other researchers have found that RNN stability and performance can be improved by using nonzero initial values (Zimmermann et al., 2006a).

The network inputs to the output units can be calculated at the same time as the hidden activations:

$$a_k^t = \sum_{h=1}^{H} w_{hk} b_h^t \tag{3.32}$$

For sequence classification and segment classification tasks (Section 2.3) the MLP output activation functions described in Section 3.1.2 (that is, logistic sigmoid for two classes and softmax for multiple classes) can be reused for RNNs, with the classification targets typically presented at the ends of the sequences or segments. It follows that the loss functions in Section 3.1.3 can be reused too. Temporal classification is more challenging, since the locations of the target classes are unknown. Chapter 7 introduces an output layer specifically designed for temporal classification with RNNs.

### 3.2.2 *Backward Pass*

Given the partial derivatives of some differentiable loss function $\mathcal{L}$ with respect to the network outputs, the next step is to determine the derivatives with respect to the weights. Two well-known algorithms have been devised to efficiently calculate weight derivatives for RNNs: real time recurrent learning (RTRL; Robinson and Fallside, 1987) and backpropagation through time (BPTT; Williams and Zipser, 1995; Werbos, 1990). We focus on BPTT since it is both conceptually simpler and more efficient in computation time (though not in memory).

Like standard backpropagation, BPTT consists of a repeated application of the chain rule. The subtlety is that, for recurrent networks, the loss function depends on the activation of the hidden layer not only through its influence on the output layer, but also through its influence on the hidden layer at the next timestep. Therefore

$$\delta_h^t = \theta'(a_h^t)\left(\sum_{k=1}^{K} \delta_k^t w_{hk} + \sum_{h'=1}^{H} \delta_{h'}^{t+1} w_{hh'}\right) \tag{3.33}$$

where

$$\delta_j^t \stackrel{\text{def}}{=} \frac{\partial \mathcal{L}}{\partial a_j^t} \tag{3.34}$$

The complete sequence of $\delta$ terms can be calculated by starting at $t = T$ and recursively applying (3.33), decrementing $t$ at each step. (Note that $\delta_j^{T+1} = 0 \; \forall j$, since no error is received from beyond the end of the sequence). Finally, bearing in mind that the same weights are reused at every timestep, we sum over the whole sequence to get the derivatives with respect to the network weights: d

$$\frac{\partial \mathcal{L}}{\partial w_{ij}} = \sum_{t=1}^{T} \frac{\partial \mathcal{L}}{\partial a_j^t} \frac{\partial a_j^t}{\partial w_{ij}} = \sum_{t=1}^{T} \delta_j^t b_i^t \tag{3.35}$$

### 3.2.3 Unfolding

A useful way to visualise RNNs is to consider the update graph formed by 'unfolding' the network along the input sequence. Figure 3.4 shows part of an unfolded RNN. Note that the unfolded graph (unlike Figure 3.3) contains no cycles; otherwise the forward and backward pass would not be well defined.

Viewing RNNs as unfolded graphs makes it easier to generalise to networks with more complex update dependencies. We will encounter such a network in the next section, and again when we consider multidimensional networks in Chapter 8 and hierarchical networks in Chapter 9.

### 3.2.4 Bidirectional Networks

For many sequence labelling tasks it is beneficial to have access to future as well as past context. For example, when classifying a particular written letter, it is helpful to know the letters coming after it as well as those before. However, since standard RNNs process sequences in temporal order, they ignore future context. An obvious solution is to add a time-window of future context to the network input. However, as well as increasing the number of input weights, this approach suffers from the same problems as

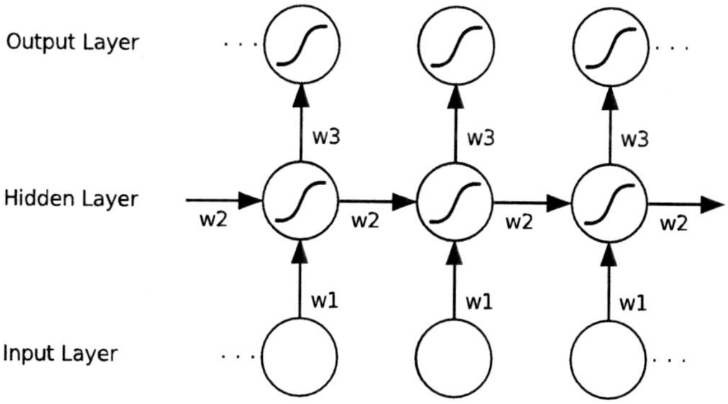


**Fig. 3.4 An unfolded recurrent network.** Each node represents a layer of network units at a single timestep. The weighted connections from the input layer to hidden layer are labelled 'w1', those from the hidden layer to itself (i.e. the recurrent weights) are labelled 'w2' and the hidden to output weights are labelled 'w3'. Note that the same weights are reused at every timestep. Bias weights are omitted for clarity.

the time-window methods discussed in Sections 2.3.1 and 2.3.2: namely intolerance of distortions, and a fixed range of context. Another possibility is to introduce a delay between the inputs and the targets, thereby giving the network a few timesteps of future context. This method retains the RNN's robustness to distortions, but it still requires the range of future context to be determined by hand. Furthermore it places an unnecessary burden on the network by forcing it to 'remember' the original input, and its previous context, throughout the delay. In any case, neither of these approaches remove the asymmetry between past and future information.

Bidirectional recurrent neural networks (BRNNs; Schuster and Paliwal, 1997; Schuster, 1999; Baldi et al., 1999) offer a more elegant solution. The basic idea of BRNNs is to present each training sequence forwards and backwards to two separate recurrent hidden layers, both of which are connected to the same output layer. This structure provides the output layer with complete past and future context for every point in the input sequence, without displacing the inputs from the relevant targets. BRNNs have previously given improved results in various domains, notably protein secondary structure prediction (Baldi et al., 2001; Chen and Chaudhari, 2004) and speech processing (Schuster, 1999; Fukada et al., 1999), and we find that they consistently outperform unidirectional RNNs at sequence labelling.

An unfolded bidirectional network is shown in Figure 3.5.

The forward pass for the BRNN hidden layers is the same as for a unidirectional RNN, except that the input sequence is presented in opposite directions to the two hidden layers, and the output layer is not updated until both hidden layers have processed the entire input sequence:

**for** $t = 1$ to $T$ **do**
 Forward pass for the forward hidden layer, storing activations at each timestep
**for** $t = T$ to 1 **do**
 Forward pass for the backward hidden layer, storing activations at each timestep
**for** all $t$, in any order **do**
 Forward pass for the output layer, using the stored activations from both hidden layers

**Algorithm 3.1** BRNN Forward Pass

Similarly, the backward pass proceeds as for a standard RNN trained with BPTT, except that all the output layer $\delta$ terms are calculated first, then fed back to the two hidden layers in opposite directions:

**for** all $t$, in any order **do**
 Backward pass for the output layer, storing $\delta$ terms at each timestep
**for** $t = T$ to 1 **do**
 BPTT backward pass for the forward hidden layer, using the stored $\delta$ terms from the output layer
**for** $t = 1$ to $T$ **do**
 BPTT backward pass for the backward hidden layer, using the stored $\delta$ terms from the output layer

**Algorithm 3.2** BRNN Backward Pass

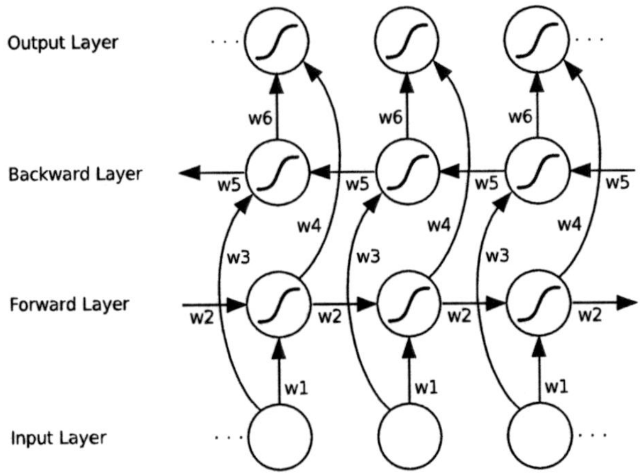


**Fig. 3.5 An unfolded bidirectional network.** Six distinct sets of weights are reused at every timestep, corresponding to the input-to-hidden, hidden-to-hidden and hidden-to-output connections of the two hidden layers. Note that no information flows between the forward and backward hidden layers; this ensures that the unfolded graph is acyclic.

#### 3.2.4.1 Causal Tasks

One objection to bidirectional networks is that they violate causality. Clearly, for tasks such as financial prediction or robot navigation, an algorithm that requires access to future inputs is unfeasible. However, there are many problems for which causality is unnecessary. Most obviously, if the input sequences are spatial and not temporal there is no reason to distinguish between past and future inputs. This is perhaps why protein structure prediction is the domain where BRNNs have been most widely adopted (Baldi et al., 2001; Thireou and Reczko, 2007). However BRNNs can also be applied to temporal tasks, as long as the network outputs are only needed at the end of some input segment. For example, in speech and handwriting recognition, the data is usually divided up into sentences, lines, or dialogue turns, each of which is completely processed before the output labelling is required. Furthermore, even for online temporal tasks, such as automatic dictation, bidirectional algorithms can be used as long as it is acceptable to wait for some natural break in the input, such as a pause in speech, before processing a section of the data.

### 3.2.5 Sequential Jacobian

It should be clear from the preceding discussions that the ability to make use of contextual information is vitally important for sequence labelling.

It therefore seems desirable to have a way of analysing exactly where and how an algorithm uses context during a particular data sequence. For RNNs, we can take a step towards this by measuring the sensitivity of the network outputs to the network inputs.

For feedforward neural networks, the *Jacobian* $J$ is the matrix of partial derivatives of the network output vector $y$ with respect to the input vector $x$:

$$J_{ki} = \frac{\partial y_k}{\partial x_i} \tag{3.36}$$

These derivatives measure the relative sensitivity of the outputs to small changes in the inputs, and can therefore be used, for example, to detect irrelevant inputs. The Jacobian can be extended to recurrent neural networks by specifying the timesteps at which the input and output variables are measured

$$J_{ki}^{tt'} = \frac{\partial y_k^t}{\partial x_i^{t'}} \tag{3.37}$$

We refer to the resulting four-dimensional matrix as the *sequential Jacobian.*

Figure 3.6 provides a sample plot of a slice through the sequential Jacobian. In general we are interested in observing the sensitivity of an output at one timestep (for example, the point when the network outputs a label) to the inputs at all timesteps in the sequence. Note that the absolute magnitude of the derivatives is not important. What matters is the *relative* magnitudes

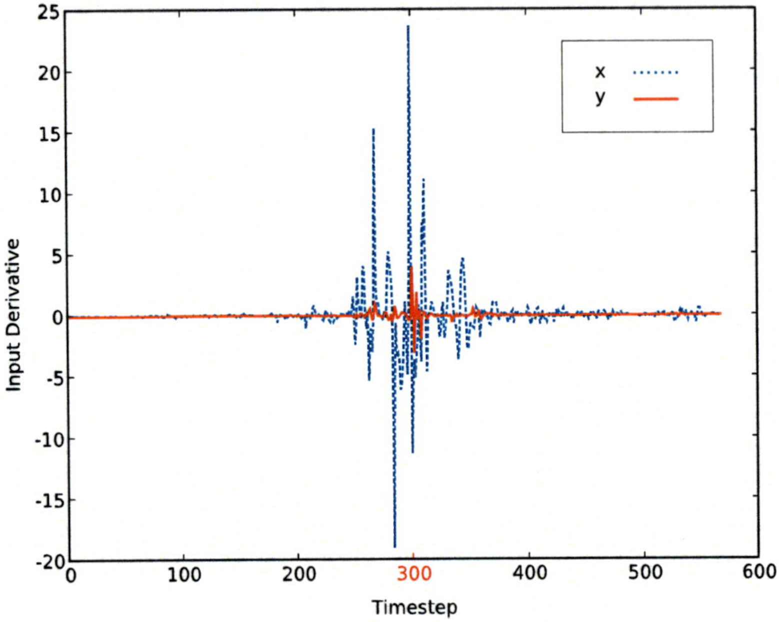


**Fig. 3.6 Sequential Jacobian for a bidirectional network during an online handwriting recognition task.** The derivatives of a single output unit at time $t = 300$ are evaluated with respect to the two inputs (corresponding to the $x$ and $y$ coordinates of the pen) at all times throughout the sequence. For bidirectional networks, the magnitude of the derivatives typically forms an 'envelope' centred on $t$. In this case the derivatives remains large for about 100 timesteps before and after $t$. The magnitudes are greater for the input corresponding to the $x$ coordinate (blue line) because this has a smaller normalised variance than the $y$ input ($x$ tends to increase steadily as the pen moves from left to right, whereas $y$ fluctuates about a fixed baseline); this does *not* imply that the network makes more use of the x coordinates than the y coordinates.

of the derivatives to each other, since this determines the relative degree to which the output is influenced by each input.

Slices like that shown in Figure 3.6 can be calculated with a simple modification of the RNN backward pass described in Section 3.2.2. First, all output delta terms are set to zero except some $\delta_k^t$, corresponding to the time $t$ and output $k$ we are interested to. This term is set equal to its own activation during the forward pass, i.e. $\delta_k^t = y_k^t$. The backward pass is then carried out as usual, and the resulting delta terms at the input layer correspond to the sensitivity of the output to the inputs over time. The intermediate delta terms (such as those in the hidden layer) are also potentially interesting, since they reveal the responsiveness of the output to different parts of the network over time.

The sequential Jacobian will be used throughout the book as a means of analysing the use of context by RNNs. However it should be stressed that sensitivity does not correspond directly to contextual importance. For example, the sensitivity may be very large towards an input that never changes, such as a corner pixel in a set of images with a fixed colour background, or the first timestep in a set of audio sequences that always begin in silence, since the network does not 'expect' to see any change there. However, this input will not provide any useful context information. Also, as shown in Figure 3.6, the sensitivity will be larger for inputs with lower variance, since the network is tuned to smaller changes. But this does not mean that these inputs are more important than those with larger variance.

## 3.3 Network Training

So far we have discussed how neural networks can be differentiated with respect to loss functions, and thereby trained with gradient descent. However, to ensure that network training is both effective and tolerably fast, and that it generalises well to unseen data, several issues must be addressed.

### *3.3.1 Gradient Descent Algorithms*

Most obviously, we need to decide how to follow the error gradient. The simplest method, known as *steepest descent* or just *gradient descent*, is to repeatedly take a small, fixed-size step in the direction of the negative error gradient of the loss function:

$$\Delta w^n = -\alpha \frac{\partial \mathcal{L}}{\partial w^n} \tag{3.38}$$

where $\Delta w^n$ is the $n^{th}$ weight update, $\alpha \in [0, 1]$ is the *learning rate* and $w^n$ is the weight vector before $\Delta w^n$ is applied. This process is repeated until some *stopping criteria* (such as failure to reduce the loss for a given number of steps) is met.

A major problem with steepest descent is that it easily gets stuck in local minima. This can be mitigated by the addition of a *momentum* term (Plaut et al., 1986), which effectively adds inertia to the motion of the algorithm through weight space, thereby speeding up convergence and helping to escape from local minima:

$$\Delta w^n = m\Delta w^{n-1} - \alpha \frac{\partial \mathcal{L}}{\partial w^n} \tag{3.39}$$

where $m \in [0, 1]$ is the momentum parameter.

When the above gradients are calculated with respect to a loss function defined over the entire training set, the weight update procedure is referred to as *batch learning*. This is in contrast to *online* or *sequential* learning, where weight updates are performed using the gradient with respect to individual training examples. Pseudocode for online learning with gradient descent is provided in Algorithm 3.3.

```
while stopping criteria not met do
  Randomise training set order
  for each example in the training set do
    Run forward and backward pass to calculate the gradient
    Update weights with gradient descent algorithm
```

**Algorithm 3.3** Online Learning with Gradient Descent

A large number of sophisticated gradient descent algorithms have been developed, such as RPROP (Riedmiller and Braun, 1993), quickprop (Fahlman, 1989), conjugate gradients (Hestenes and Stiefel, 1952; Shewchuk, 1994) and L-BFGS (Byrd et al., 1995), that generally outperform steepest descent at batch learning. However steepest descent is much better suited than they are to online learning, because it takes very small steps at each weight update and can therefore tolerate constantly changing gradients.

Online learning tends to be more efficient than batch learning when large datasets containing significant redundancy or regularity are used (LeCun et al., 1998c). In addition, the stochasticity of online learning can help to escape from local minima (LeCun et al., 1998c), since the loss function is different for each training example. The stochasticity can be further increased by randomising the order of the sequences in the training set before each pass through the training set (often referred to as a training *epoch*). Training set randomisation is used for all the experiments in this book.

A recently proposed alternative for online learning is *stochastic meta-descent* (Schraudolph, 2002), which has been shown to give faster convergence and improved results for a variety of neural network tasks. However our attempts to train RNNs with stochastic meta-descent were unsuccessful, and all experiments in this book were carried out using online steepest descent with momentum.

### 3.3.2 Generalisation

Although the loss functions for network training are, of necessity, defined on the training set, the real goal is to optimise performance on a test set of previously unseen data. The issue of whether training set performance carries over to the test set is referred to as *generalisation*, and is of fundamental importance to machine learning (see e.g. Vapnik, 1995; Bishop, 2006). In general the larger the training set the better the generalisation. Many methods for improved generalisation with a fixed size training set (often referred to as *regularisers*) have been proposed over the years. In this book, however, only three simple regularisers are used: early stopping, input noise and weight noise.

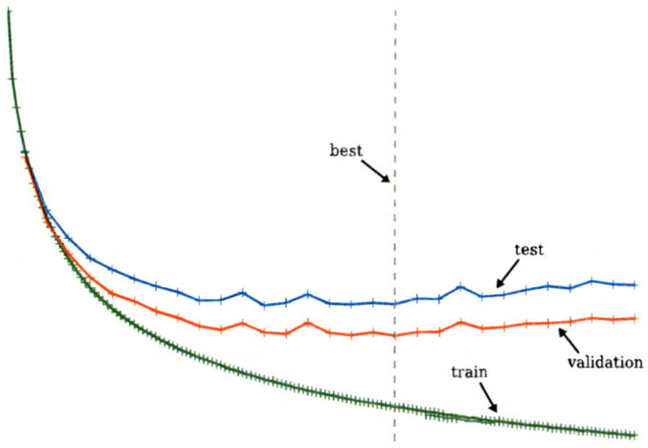


**Fig. 3.7 Overfitting on training data.** Initially, network error decreases rapidly on all datasets. Soon however it begins to level off and gradually rise on the validation and test sets. The dashed line indicates the point of best performance on the validation set, which is close, but not identical to the optimal point for the test set.

#### 3.3.2.1 Early Stopping

For early stopping, part of the training set is removed for use as a *validation set*. All stopping criteria are then tested on the validation set instead of the training set. The 'best' weight values are also chosen using the validation set, typically by picking the weights that minimise on the validation set the error function used to assess performance on the test set. In practice the two are usually done in tandem, with the error evaluated at regular intervals on the validation set, and training stopped after the error fails to decrease for a certain number of evaluations.

The test set should not be used to decide when to stop training or to choose the best weight values; these are indirect forms of training on the test set. In principle, the network should not be evaluated on the test set at all until training is complete.

During training, the error typically decreases at first on all sets, but after a certain point it begins to rise on the test and validation sets, while continuing to decrease on the training set. This behaviour, known as *overfitting*, is illustrated in Figure 3.7.

Early stopping is perhaps the simplest and most universally applicable method for improved generalisation. However one drawback is that some of the training set has to be sacrificed for the validation set, which can lead to reduced performance, especially if the training set is small. Another problem is that there is no way of determining *a priori* how big the validation set should be. For the experiments in this book, we typically use five to ten percent of the training set for validation. Note that the validation set does not have to be an accurate predictor of test set performance; it is only important that overfitting begins at approximately the same time on both of them.

#### 3.3.2.2 Input Noise

Adding zero-mean, fixed-variance Gaussian noise to the network inputs during training (sometimes referred to as *training with jitter*) is a well-established method for improved generalisation (An, 1996; Koistinen and Holmström, 1991; Bishop, 1995). The desired effect is to artificially enhance the size of the training set, and thereby improve generalisation, by generating new inputs with the same targets as the original ones.

One problem with input noise is that it is difficult to determine in advance how large the noise variance should be. Although various rules of thumb exist, the most reliable method is to set the variance empirically on the validation set.

A more fundamental difficulty is that input perturbations are only effective if they reflect the variations found in the real data. For example, adding Gaussian noise to individual pixel values in an image will not generate a substantially different image (only a 'speckled' version of the original) and is therefore unlikely to aid generalisation to new images. Independently perturbing the points in a smooth trajectory is ineffectual for the same reason. Input perturbations tailored towards a particular dataset have been shown to be highly effective at improving generalisation (Simard et al., 2003); however this requires a prior model of the data variations, which is not usually available.

Figure 3.8 illustrates the distinction between Gaussian input noise and data-specific input perturbations.

Input noise should be regenerated for every example presented to the network during training; in particular, the same noise should not be re-used for a given example as the network cycles through the data. Input noise should not be added during testing, as doing so will hamper performance.

#### 3.3.2.3 Weight Noise

An alternative regularisation strategy is to add zero-mean, fixed variance Gaussian noise to the network *weights* (Murray and Edwards, 1994; Jim et al., 1996). Because *weight noise* or *synaptic noise* acts on the network's internal representation of the inputs, rather than the inputs themselves, it can be used for any data type. However weight noise is typically less effective than carefully designed input perturbations, and can lead to very slow convergence.

Weight noise can be used to 'simplify' neural networks, in the sense of reducing the amount of information required to transmit the network (Hinton and van Camp, 1993). Intuitively this is because noise reduces the precision with which the weights must be described. Simpler networks are preferable because they tend to generalise better—a manifestation of Occam's razor.

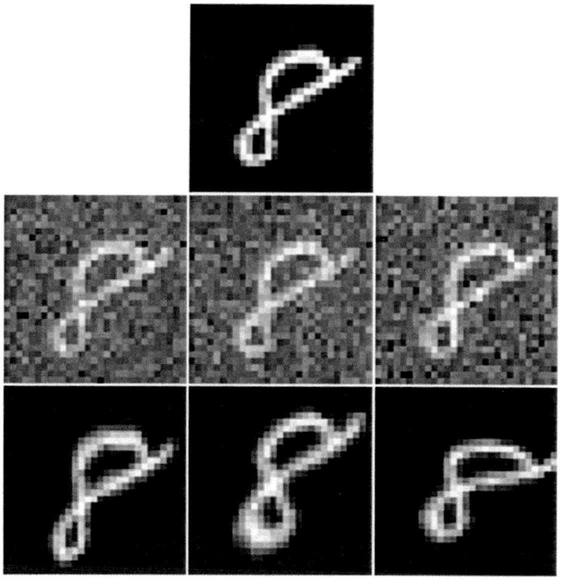

**Fig. 3.8 Different Kinds of Input Perturbation.** A handwritten digit from the MNIST database (top) is shown perturbed with Gaussian noise (centre) and elastic deformations (bottom). Since Gaussian noise does not alter the outline of the digit and the noisy images all look qualitatively the same, this approach is unlikely to improve generalisation on MNIST. The elastic distortions, on the other hand, appear to create different handwriting samples out of the same image, and can therefore be used to artificially extend the training set.

Algorithm 3.4 shows how weight noise should be applied during online learning with gradient descent.

**while** stopping criteria not met **do**
  Randomise training set order
  **for** each example in the training set **do**
    Add zero mean Gaussian noise to weights
    Run forward and backward pass to calculate the gradient
    Restore original weights
    Update weights with gradient descent algorithm

**Algorithm 3.4** Online Learning with Gradient Descent and Weight Noise

As with input noise, weight noise should not be added when the network is evaluated on test data.

### 3.3.3 Input Representation

Choosing a suitable representation of the input data is a vital part of any machine learning task. Indeed, in some cases it is more important to the final performance than the algorithm itself. Neural networks, however, tend to be relatively robust to the choice of input representation: for example, in previous work on phoneme recognition, RNNs were shown to perform almost equally well using a wide range of speech preprocessing methods (Robinson et al., 1990). We report similar findings in Chapters 7 and 9, with very different input representations found to give roughly equal performance for both speech and handwriting recognition.

The only requirements for neural network input representations are that they are complete (in the sense of containing all information required to successfully predict the outputs) and reasonably compact. Although irrelevant inputs are not as much of a problem for neural networks as they are for algorithms suffering from the so-called *curse of dimensionality* (see e.g. Bishop, 2006), having a very high dimensional input space leads to an excessive number of input weights and poor generalisation. Beyond that the choice of input representation is something of a black art, whose aim is to make the relationship between the inputs and the targets as simple as possible.

One procedure that should be carried out for all neural network input data is to *standardise* the components of the input vectors to have mean 0 and standard deviation 1 over the training set. That is, first calculate the mean

$$m_i = \frac{1}{|S|} \sum_{x \in S} x_i \tag{3.40}$$

and standard deviation

$$\sigma_i = \sqrt{\frac{1}{|S|} \sum_{x \in S} (x_i - m_i)^2} \tag{3.41}$$

of each component of the input vector, then calculate the standardised input vectors $\hat{x}$ using

$$\hat{x}_i = \frac{x_i - m_i}{\sigma_i} \tag{3.42}$$

This procedure does not alter the information in the training set, but it improves performance by putting the input values in a range more suitable for the standard activation functions (LeCun et al., 1998c). Note that the test and validation sets should be standardised with the mean and standard deviation of the training set.

Input standardisation can have a huge effect on network performance, and was carried out for all the experiments in this book.

### 3.3.4 Weight Initialisation

Many gradient descent algorithms for neural networks require small, random, initial values for the weights. For the experiments in this book, we initialised the weights with either a flat random distribution in the range $[-0.1, 0.1]$ or a Gaussian distribution with mean 0, standard deviation 0.1. However, we did not find our results to be very sensitive to either the distribution or the range. A consequence of having random initial conditions is that each experiment must be repeated several times to determine significance.

# Chapter 4
# Long Short-Term Memory

As discussed in the previous chapter, an important benefit of recurrent neural networks is their ability to use contextual information when mapping between input and output sequences. Unfortunately, for standard RNN architectures, the range of context that can be in practice accessed is quite limited. The problem is that the influence of a given input on the hidden layer, and therefore on the network output, either decays or blows up exponentially as it cycles around the network's recurrent connections. This effect is often referred to in the literature as the *vanishing gradient problem* (Hochreiter, 1991; Hochreiter et al., 2001a; Bengio et al., 1994). The vanishing gradient problem is illustrated schematically in Figure 4.1

Numerous attempts were made in the 1990s to address the problem of vanishing gradients for RNNs. These included non-gradient based training algorithms, such as simulated annealing and discrete error propagation (Bengio et al., 1994), explicitly introduced time delays (Lang et al., 1990; Lin et al., 1996; Plate, 1993) or time constants (Mozer, 1992), and hierarchical sequence compression (Schmidhuber, 1992). The approach favoured by this book is the *Long Short-Term Memory* (LSTM) architecture (Hochreiter and Schmidhuber, 1997).

This chapter reviews the background material for LSTM. Section 4.1 describes the basic structure of LSTM and explains how it tackles the vanishing gradient problem. Section 4.3 discusses an approximate and an exact algorithm for calculating the LSTM error gradient. Section 4.4 describes some enhancements to the basic LSTM architecture. Section 4.2 discusses the effect of preprocessing on long range dependencies. Section 4.6 provides all the equations required to train and apply LSTM networks.

## 4.1 Network Architecture

The LSTM architecture consists of a set of recurrently connected subnets, known as memory blocks. These blocks can be thought of as a differentiable version of the memory chips in a digital computer. Each block contains one or more self-connected memory cells and three multiplicative units—the input,

A. Graves: Supervised Sequence Labell. with Recur. Neur. Networks, SCI 385, pp. 37–45.
springerlink.com 

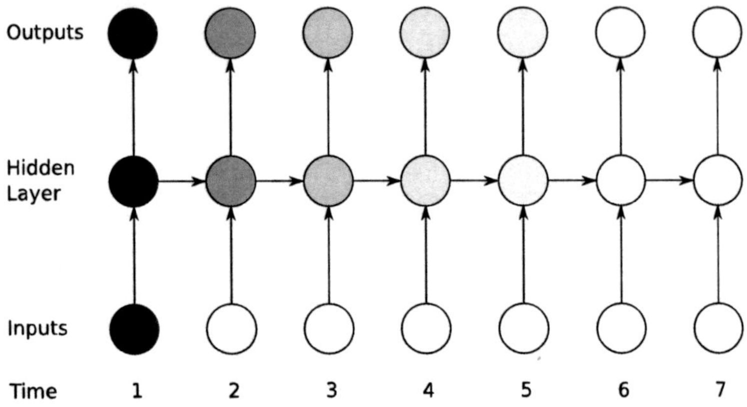


**Fig. 4.1 The vanishing gradient problem for RNNs.** The shading of the nodes in the unfolded network indicates their sensitivity to the inputs at time one (the darker the shade, the greater the sensitivity). The sensitivity decays over time as new inputs overwrite the activations of the hidden layer, and the network 'forgets' the first inputs.

output and forget gates—that provide continuous analogues of write, read and reset operations for the cells.

Figure 4.2 provides an illustration of an LSTM memory block with a single cell. An LSTM network is the same as a standard RNN, except that the summation units in the hidden layer are replaced by memory blocks, as illustrated in Fig. 4.3. LSTM blocks can also be mixed with ordinary summation units, although this is typically not necessary. The same output layers can be used for LSTM networks as for standard RNNs.

The multiplicative gates allow LSTM memory cells to store and access information over long periods of time, thereby mitigating the vanishing gradient problem. For example, as long as the input gate remains closed (i.e. has an activation near 0), the activation of the cell will not be overwritten by the new inputs arriving in the network, and can therefore be made available to the net much later in the sequence, by opening the output gate. The preservation over time of gradient information by LSTM is illustrated in Figure 4.4.

Over the past decade, LSTM has proved successful at a range of synthetic tasks requiring long range memory, including learning context free languages (Gers and Schmidhuber, 2001), recalling high precision real numbers over extended noisy sequences (Hochreiter and Schmidhuber, 1997) and various tasks requiring precise timing and counting (Gers et al., 2002). In particular, it has solved several artificial problems that remain impossible with any other RNN architecture.

Additionally, LSTM has been applied to various real-world problems, such as protein secondary structure prediction (Hochreiter et al., 2007; Chen and Chaudhari, 2005), music generation (Eck and Schmidhuber, 2002),

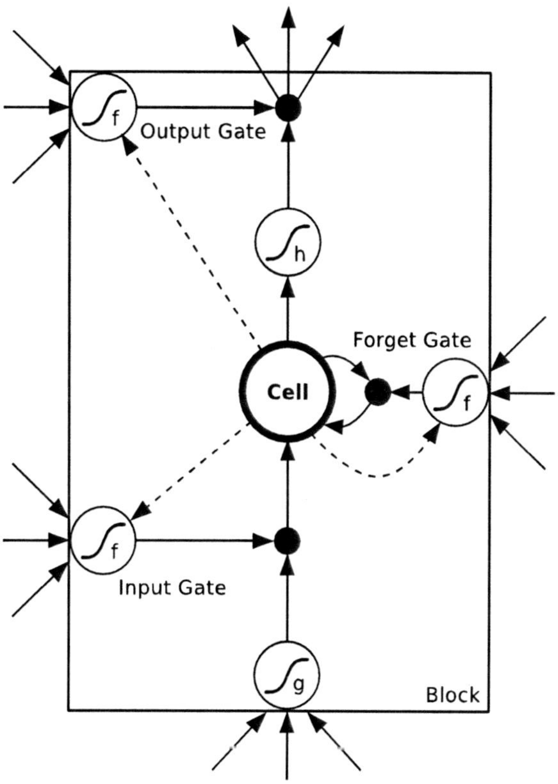


**Fig. 4.2 LSTM memory block with one cell.** The three gates are nonlinear summation units that collect activations from inside and outside the block, and control the activation of the cell via multiplications (small black circles). The input and output gates multiply the input and output of the cell while the forget gate multiplies the cell's previous state. No activation function is applied within the cell. The gate activation function 'f' is usually the logistic sigmoid, so that the gate activations are between 0 (gate closed) and 1 (gate open). The cell input and output activation functions ('g' and 'h') are usually tanh or logistic sigmoid, though in some cases 'h' is the identity function. The weighted 'peephole' connections from the cell to the gates are shown with dashed lines. All other connections within the block are unweighted (or equivalently, have a fixed weight of 1.0). The only outputs from the block to the rest of the network emanate from the output gate multiplication.

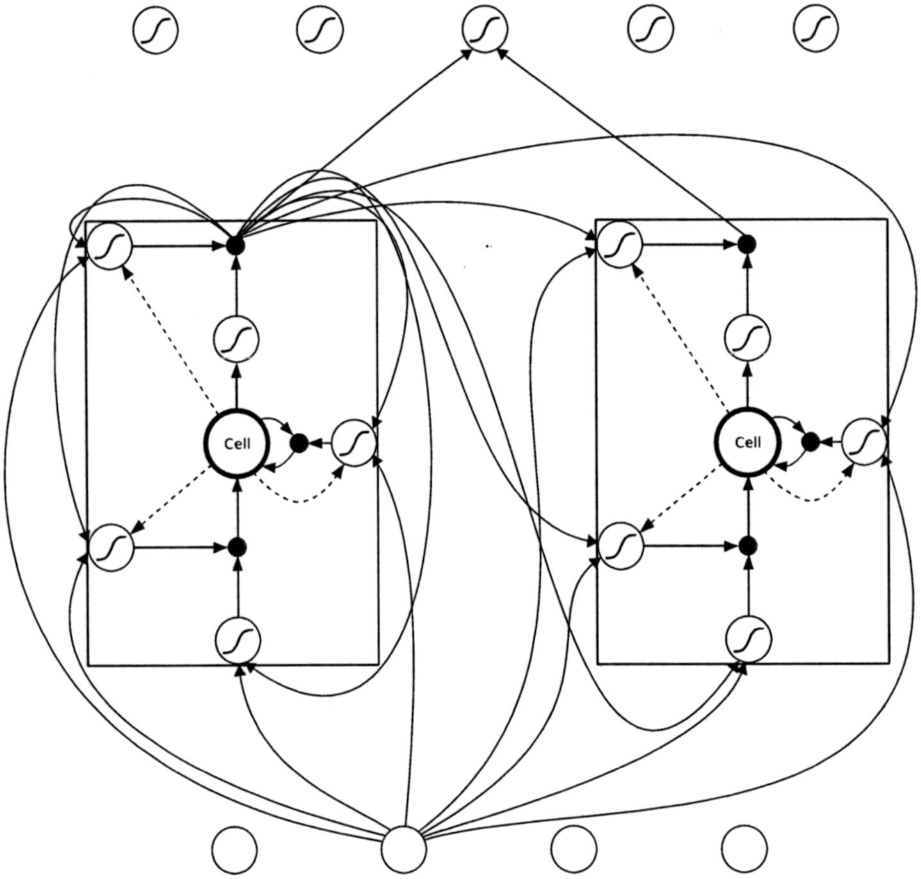


**Fig. 4.3 An LSTM network.** The network consists of four input units, a hidden layer of two single-cell LSTM memory blocks and five output units. Not all connections are shown. Note that each block has four inputs but only one output.

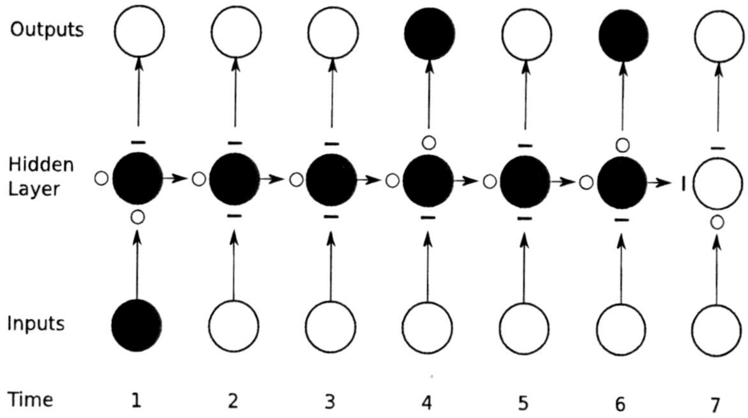


**Fig. 4.4 Preservation of gradient information by LSTM.** As in Figure 4.1 the shading of the nodes indicates their sensitivity to the inputs at time one; in this case the black nodes are maximally sensitive and the white nodes are entirely insensitive. The state of the input, forget, and output gates are displayed below, to the left and above the hidden layer respectively. For simplicity, all gates are either entirely open ('O') or closed ('—'). The memory cell 'remembers' the first input as long as the forget gate is open and the input gate is closed. The sensitivity of the output layer can be switched on and off by the output gate without affecting the cell.

reinforcement learning (Bakker, 2002), speech recognition (Graves and Schmidhuber, 2005b; Graves et al., 2006) and handwriting recognition (Liwicki et al., 2007; Graves et al., 2008). As would be expected, its advantages are most pronounced for problems requiring the use of long range contextual information.

## 4.2 Influence of Preprocessing

The above discussion raises an important point about the influence of preprocessing. If we can find a way to transform a task containing long range contextual dependencies into one containing only short-range dependencies before presenting it to a sequence learning algorithm, then architectures such as LSTM become somewhat redundant. For example, a raw speech signal typically has a sampling rate of over 40 kHz. Clearly, a great many timesteps would have to be spanned by a sequence learning algorithm attempting to label or model an utterance presented in this form. However when the signal is first transformed into a 100 Hz series of mel-frequency cepstral coefficients, it becomes feasible to model the data using an algorithm whose contextual range is relatively short, such as a hidden Markov model.

Nonetheless, if such a transform is difficult or unknown, or if we simply wish to get a good result without having to design task-specific preprocessing methods, algorithms capable of handling long time dependencies are essential.

## 4.3 Gradient Calculation

Like the networks discussed in the last chapter, LSTM is a differentiable function approximator that is typically trained with gradient descent. Recently, non gradient-based training methods of LSTM have also been considered (Wierstra et al., 2005; Schmidhuber et al., 2007), but they are outside the scope of this book.

The original LSTM training algorithm (Hochreiter and Schmidhuber, 1997) used an approximate error gradient calculated with a combination of Real Time Recurrent Learning (RTRL; Robinson and Fallside, 1987) and Backpropagation Through Time (BPTT; Williams and Zipser, 1995). The BPTT part was truncated after one timestep, because it was felt that long time dependencies would be dealt with by the memory blocks, and not by the (vanishing) flow of activation around the recurrent connections. Truncating the gradient has the benefit of making the algorithm completely online, in the sense that weight updates can be made after every timestep. This is an important property for tasks such as continuous control or time-series prediction.

However, it is also possible to calculate the exact LSTM gradient with untruncated BPTT (Graves and Schmidhuber, 2005b). As well as being more accurate than the truncated gradient, the exact gradient has the advantage of being easier to debug, since it can be checked numerically using the technique described in Section 3.1.4.1. Only the exact gradient is used in this book, and the equations for it are provided in Section 4.6.

## 4.4 Architectural Variants

In its original form, LSTM contained only input and output gates. The forget gates (Gers et al., 2000), along with additional peephole weights (Gers et al., 2002) connecting the gates to the memory cell were added later to give *extended LSTM* (Gers, 2001). The purpose of the forget gates was to provide a way for the memory cells to reset themselves, which proved important for tasks that required the network to 'forget' previous inputs. The peephole connections, meanwhile, improved the LSTM's ability to learn tasks that require precise timing and counting of the internal states.

Since LSTM is entirely composed of simple multiplication and summation units, and connections between them, it is straightforward to create further variants of the block architecture. Indeed it has been shown that alternative structures with equally good performance on toy problems such as learning context-free and context-sensitive languages can be evolved automatically (Bayer et al., 2009). However the standard extended form appears to be a good general purpose structure for sequence labelling, and is used exclusively in this book.

## 4.5 Bidirectional Long Short-Term Memory

Using LSTM as the network architecture in a bidirectional recurrent neural network (Section 3.2.4) yields bidirectional LSTM (Graves and Schmidhuber, 2005a,b; Chen and Chaudhari, 2005; Thireou and Reczko, 2007). Bidirectional LSTM provides access to long range context in both input directions, and will be used extensively in later chapters.

## 4.6 Network Equations

This section provides the equations for the activation (forward pass) and BPTT gradient calculation (backward pass) of an LSTM hidden layer within a recurrent neural network.

As before, $w_{ij}$ is the weight of the connection from unit $i$ to unit $j$, the network input to unit $j$ at time $t$ is denoted $a_j^t$ and activation of unit $j$ at time $t$ is $b_j^t$. The LSTM equations are given for a single memory block only. For multiple blocks the calculations are simply repeated for each block, in any order. The subscripts $\iota$, $\phi$ and $\omega$ refer respectively to the input gate, forget gate and output gate of the block. The subscripts $c$ refers to one of the $C$ memory cells. The *peephole weights* from cell $c$ to the input, forget and output gates are denoted $w_{c\iota}$, $w_{c\phi}$ and $w_{c\omega}$ respectively. $s_c^t$ is the *state* of cell $c$ at time $t$ (i.e. the activation of the linear cell unit). $f$ is the activation function of the gates, and $g$ and $h$ are respectively the cell input and output activation functions.

Let $I$ be the number of inputs, $K$ be the number of outputs and $H$ be the number of cells in the hidden layer. Note that only the *cell outputs* $b_c^t$ are connected to the other blocks in the layer. The other LSTM activations, such as the states, the cell inputs, or the gate activations, are only visible within the block. We use the index $h$ to refer to cell outputs from other blocks in the hidden layer, exactly as for standard hidden units. As with standard RNNs the forward pass is calculated for a length $T$ input sequence $\mathbf{x}$ by starting at $t = 1$ and recursively applying the update equations while incrementing $t$, and the BPTT backward pass is calculated by starting at $t = T$, and recursively calculating the unit derivatives while decrementing $t$ to one (see Section 3.2 for details). The final weight derivatives are found by summing over the derivatives at each timestep, as expressed in Eqn. (3.35). Recall that

$$\delta_j^t \stackrel{\text{def}}{=} \frac{\partial \mathcal{L}}{\partial a_j^t} \tag{4.1}$$

where $\mathcal{L}$ is the loss function used for training.

The order in which the equations are calculated during the forward and backward passes is important, and should proceed as specified below. As with standard RNNs, all states and activations are initialised to zero at $t = 0$, and all $\delta$ terms are zero at $t = T + 1$.

### 4.6.1 Forward Pass

*Input Gates*

$$a_\iota^t = \sum_{i=1}^{I} w_{i\iota} x_i^t + \sum_{h=1}^{H} w_{h\iota} b_h^{t-1} + \sum_{c=1}^{C} w_{c\iota} s_c^{t-1} \tag{4.2}$$

$$b_\iota^t = f(a_\iota^t) \tag{4.3}$$

*Forget Gates*

$$a_\phi^t = \sum_{i=1}^{I} w_{i\phi} x_i^t + \sum_{h=1}^{H} w_{h\phi} b_h^{t-1} + \sum_{c=1}^{C} w_{c\phi} s_c^{t-1} \tag{4.4}$$

$$b_\phi^t = f(a_\phi^t) \tag{4.5}$$

*Cells*

$$a_c^t = \sum_{i=1}^{I} w_{ic} x_i^t + \sum_{h=1}^{H} w_{hc} b_h^{t-1} \tag{4.6}$$

$$s_c^t = b_\phi^t s_c^{t-1} + b_\iota^t g(a_c^t) \tag{4.7}$$

*Output Gates*

$$a_\omega^t = \sum_{i=1}^{I} w_{i\omega} x_i^t + \sum_{h=1}^{H} w_{h\omega} b_h^{t-1} + \sum_{c=1}^{C} w_{c\omega} s_c^t \tag{4.8}$$

$$b_\omega^t = f(a_\omega^t) \tag{4.9}$$

*Cell Outputs*

$$b_c^t = b_\omega^t h(s_c^t) \tag{4.10}$$

### 4.6.2 Backward Pass

$$\epsilon_c^t \stackrel{\text{def}}{=} \frac{\partial \mathcal{L}}{\partial b_c^t} \qquad \epsilon_s^t \stackrel{\text{def}}{=} \frac{\partial \mathcal{L}}{\partial s_c^t}$$

*Cell Outputs*

$$\epsilon_c^t = \sum_{k=1}^{K} w_{ck} \delta_k^t + \sum_{h=1}^{H} w_{ch} \delta_h^{t+1} \tag{4.11}$$

*Output Gates*

$$\delta_\omega^t = f'(a_\omega^t) \sum_{c=1}^{C} h(s_c^t) \epsilon_c^t \tag{4.12}$$

*States*

$$\epsilon_s^t = b_\omega^t h'(s_c^t)\epsilon_c^t + b_\phi^{t+1}\epsilon_s^{t+1} + w_{c\iota}\delta_\iota^{t+1} + w_{c\phi}\delta_\phi^{t+1} + w_{c\omega}\delta_\omega^t \tag{4.13}$$

*Cells*

$$\delta_c^t = b_\iota^t g'(a_c^t)\epsilon_s^t \tag{4.14}$$

*Forget Gates*

$$\delta_\phi^t = f'(a_\phi^t)\sum_{c=1}^{C} s_c^{t-1}\epsilon_s^t \tag{4.15}$$

*Input Gates*

$$\delta_\iota^t = f'(a_\iota^t)\sum_{c=1}^{C} g(a_c^t)\epsilon_s^t \tag{4.16}$$

# Chapter 5
# A Comparison of Network Architectures

This chapter presents an experimental comparison between various neural network architectures on a framewise phoneme classification task (Graves and Schmidhuber, 2005a,b). Framewise phoneme classification is an example of a segment classification task (see Section 2.3.2). It tests an algorithm's ability to segment and recognise the constituent parts of a speech signal, requires the use of contextual information, and can be regarded as a first step to continuous speech recognition.

Context is of particular importance in speech recognition due to phenomena such as *co-articulation*, where the human articulatory system blurs together adjacent sounds in order to produce them rapidly and smoothly. In many cases it is difficult to identify a particular phoneme without knowing the phonemes that occur before and after it. The main conclusion of this chapter is that network architectures capable of accessing more context give better performance in phoneme classification, and are therefore more suitable for speech recognition.

Section 5.1 describes the experimental data and task. Section 5.2 gives an overview of the various neural network architectures and Section 5.3 describes how they are trained, while Section 5.4 presents the experimental results.

## 5.1 Experimental Setup

The data for the experiments came from the TIMIT corpus (Garofolo et al., 1993) of prompted speech, collected by Texas Instruments. The utterances in TIMIT were chosen to be phonetically rich, and the speakers represent a wide variety of American dialects. The audio data is divided into sentences, each of which is accompanied by a phonetic transcript.

The task was to classify every input timestep, or *frame* in audio parlance, according to the phoneme it belonged to. For consistency with the literature, we used the complete set of 61 phonemes provided in the transcriptions. In continuous speech recognition, it is common practice to use a reduced set of

A. Graves: Supervised Sequence Labell. with Recur. Neur. Networks, SCI 385, pp. 47–56.
springerlink.com 

phonemes (Robinson, 1991), by merging those with similar sounds, and not separating closures from stops.

The standard TIMIT corpus comes partitioned into training and test sets, containing 3,696 and 1,344 utterances respectively. In total there were 1,124,823 frames in the training set, and 410,920 in the test set. No speakers or sentences exist in both the training and test sets. 184 of the training set utterances (chosen randomly, but kept constant for all experiments) were used as a validation set for early stopping. All results for the training and test sets were recorded at the point of lowest error on the validation set.

The following preprocessing, which is standard in speech recognition was used for the audio data. The input data was characterised as a sequence of vectors of 26 coefficients, consisting of twelve Mel-frequency cepstral coefficients (MFCC) plus energy and first derivatives of these magnitudes. First the coefficients were computed every 10ms over 25ms windows. Then a Hamming window was applied, a Mel-frequency filter bank of 26 channels was computed and, finally, the MFCC coefficients were calculated with a 0.97 pre-emphasis coefficient. The preprocessing was carried out using the *Hidden Markov Model Toolkit* (Young et al., 2006).

## 5.2 Network Architectures

We used the following five neural network architectures in our experiments (henceforth referred to by the abbreviations in brackets):

- Bidirectional LSTM, with two hidden LSTM layers (forwards and backwards), both containing 93 memory blocks of one cell each (BLSTM)
- Unidirectional LSTM, with one hidden LSTM layer, containing 140 one-cell memory blocks, trained backwards with no target delay, and forwards with delays from 0 to 10 frames (LSTM)
- Bidirectional RNN with two hidden layers containing 185 sigmoid units each (BRNN)
- Unidirectional RNN with one hidden layer containing 275 sigmoid units, trained with target delays from 0 to 10 frames (RNN)
- MLP with one hidden layer containing 250 sigmoid units, and symmetrical time-windows from 0 to 10 frames (MLP)

The hidden layer sizes were chosen to ensure that all networks had roughly the same number of weights $W$ ($\approx 100,000$), thereby providing a fair comparison. Note however that for the MLPs the number of weights grew with the time-window size, and $W$ ranged from 22,061 to 152,061. All networks contained an input layer of size 26 (one for each MFCC coefficient), and an output layer of size 61 (one for each phoneme). The input layers were fully connected to the hidden layers and the hidden layers were fully connected to the output layers. For the recurrent networks, the hidden layers were also fully connected to themselves. The LSTM blocks had the following activation functions: logistic sigmoids in the range $[-2, 2]$ for the input and output activation functions of the cell ($g$ and $h$ in Figure 4.2), and in the range $[0, 1]$ for the gates. The non-LSTM networks had logistic sigmoid activations in the range $[0, 1]$ in the hidden layers. All units were biased.

Figure 5.1 illustrates the behaviour of the different architectures during classification.

### 5.2.1 Computational Complexity

For all networks, the computational complexity was dominated by the $\mathcal{O}(W)$ feedforward and feedback operations. This means that the bidirectional networks and the LSTM networks did not take significantly more time per training epoch than the unidirectional or RNN or (equivalently sized) MLP networks.

### 5.2.2 Range of Context

Only the bidirectional networks had access to the complete context of the frame being classified (i.e. the whole input sequence). For MLPs, the amount of context depended on the size of the time-window. The results for the MLP with no time-window (presented only with the current frame) give a baseline for performance without context information. However, some context is implicitly present in the window averaging and first-derivatives included in the preprocessor.

Similarly, for unidirectional LSTM and RNN, the amount of future context depended on the size of target delay. The results with no target delay (trained forwards or backwards) give a baseline for performance with context in one direction only.

### 5.2.3 Output Layers

For the output layers, we used the cross entropy error function and the softmax activation function, as discussed in Sections 3.1.2 and 3.1.3. The softmax function ensures that the network outputs are all between zero and one, and that they sum to one on every timestep. This means they can be interpreted as the posterior probabilities of the phonemes at a given frame, given all the inputs up to the current one (with unidirectional networks) or all the inputs in the whole sequence (with bidirectional networks).

Several alternative error functions have been studied for this task (Chen and Jamieson, 1996). One modification in particular has been shown to have a positive effect on continuous speech recognition. This is to weight the error according to the duration of the current phoneme, ensuring that short

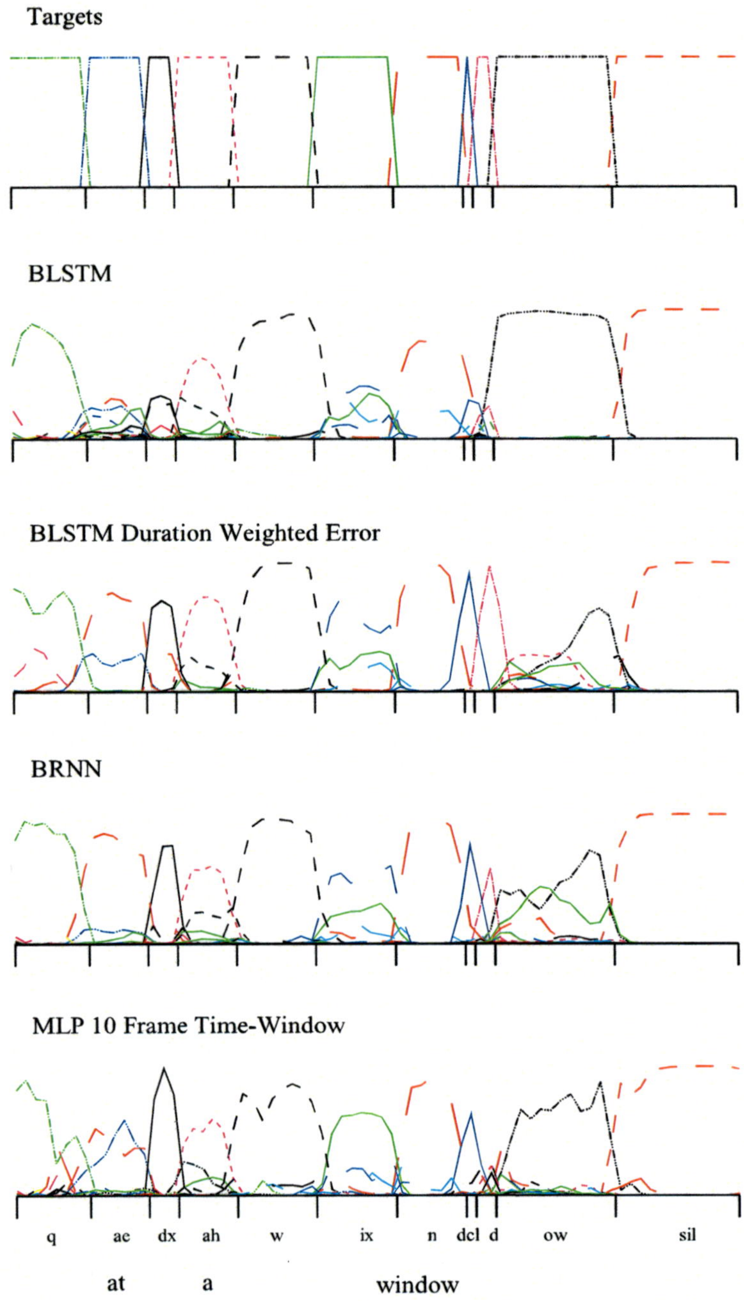


**Fig. 5.1 Various networks classifying the excerpt "at a window" from TIMIT.** In general, the networks found the vowels more difficult than the consonants, which in English are more distinct. Adding duration weighted error to BLSTM tends to give better results on short phonemes, (e.g. the closure and stop 'dcl' and 'd'), and worse results on longer ones ('ow'), as expected. Note the more jagged trajectories for the MLP; this is a consequence of not having a recurrent hidden layer, and therefore calculating each output independently of the others.

phonemes are as significant to training as longer ones. We will return to the issue of weighted errors in the next chapter.

## 5.3 Network Training

For all architectures, we calculated the full error gradient using BPTT for each utterance, and trained the weights using online steepest descent with momentum. The same training parameters were used for all experiments: initial weights chosen from a flat random distribution with range $[-0.1, 0.1]$, a learning rate of $10^{-5}$ and a momentum of 0.9. Weight updates were carried out at the end of each sequence and the order of the training set was randomised at the start of each training epoch.

Keeping the training algorithm and parameters constant allowed us to concentrate on the effect of varying the architecture. However it is possible that different training methods would be better suited to different networks.

Note that, other than early stopping, no techniques for improved generalisation were used. It is likely the addition of either input noise (Section 3.3.2.2) or weight noise (Section 3.3.2.3) would have lead to better performance.

### *5.3.1 Retraining*

For the experiments with varied time-windows or target delays, we iteratively retrained the networks, instead of starting again from scratch. For example, for LSTM with a target delay of 2, we first trained with delay 0, then took the best network and retrained it (without resetting the weights) with delay 1, then retrained again with delay 2. To find the best networks, we retrained the LSTM networks for 5 epochs at each iteration, the RNN networks for 10, and the MLPs for 20. It is possible that longer retraining times would have given improved results. For the retrained MLPs, we had to add extra (randomised) weights from the input layers, since the input size grew with the time-window.

Although primarily a means to reduce training time, we have also found that retraining improves final performance (Graves et al., 2005a; Beringer, 2004). Indeed, the best result in this chapter was achieved by retraining (on the BLSTM network trained with a weighted error function, then retrained with normal cross-entropy error). The benefits presumably come from escaping the local minima that gradient descent algorithms tend to get caught in.

The ability of neural networks to benefit from this kind of retraining touches on the more general issue of transferring knowledge between different tasks (usually known as *transfer learning* or *meta-learning*) which has been widely studied in the neural network and general machine learning literature (see e.g. Giraud-Carrier et al., 2004).

**Table 5.1 Framewise phoneme classification results on TIMIT.** The error measure is the frame error rate (percentage of misclassified frames). BLSTM results are means over seven runs ± standard error.

| Network | Train Error (%) | Test Error (%) | Epochs |
|---|---|---|---|
| MLP (no window) | 46.4 | 48.6 | 835 |
| MLP (10 frame window) | 32.4 | 36.9 | 990 |
| RNN (delay 0) | 30.1 | 35.5 | 120 |
| LSTM (delay 0) | 29.1 | 35.4 | 15 |
| LSTM (backwards, delay 0) | 29.9 | 35.3 | 15 |
| RNN (delay 3) | 29.0 | 34.8 | 140 |
| LSTM (delay 5) | 22.4 | 34.0 | 35 |
| BLSTM (Weighted Error) | 24.3 | 31.1 | 15 |
| BRNN | 24.0 | 31.0 | 170 |
| BLSTM | 22.6±0.2 | 30.2±0.1 | 20.1±0.5 |
| BLSTM (retrained) | 21.4 | 29.8 | 17 |

## 5.4 Results

Table 5.1 summarises the performance of the different network architectures. For the MLP, RNN and LSTM networks we give both the best results, and those achieved with least contextual information (i.e. with no target delay or time-window). The complete set of results is presented in Figure 5.2.

The most obvious difference between LSTM and the RNN and MLP networks was the number of epochs required for training, as shown in Figure 5.3. In particular, BRNN took more than eight times as long to converge as BLSTM, despite having more or less equal computational complexity per timestep (see Section 5.2.1). There was a similar time increase between the unidirectional LSTM and RNN networks, and the MLPs were slower still (990 epochs for the best MLP result). A possible explanation for this is that the MLPs and RNNs require more fine-tuning of their weights to access long range contextual information.

As well as being faster, the LSTM networks were also slightly more accurate. However, the final difference in score between BLSTM and BRNN on this task is quite small (0.8%). The fact that the difference is not larger could mean that very long time dependencies are not required for this task.

It is interesting to note how much more prone to overfitting LSTM was than standard RNNs. For LSTM, after only fifteen to twenty epochs the performance on the validation and test sets would begin to fall, while that on the training set would continue to rise (the highest score we recorded on the training set with BLSTM was 86.4%). With the RNNs on the other hand, we never observed a large drop in test set score. This suggests a difference

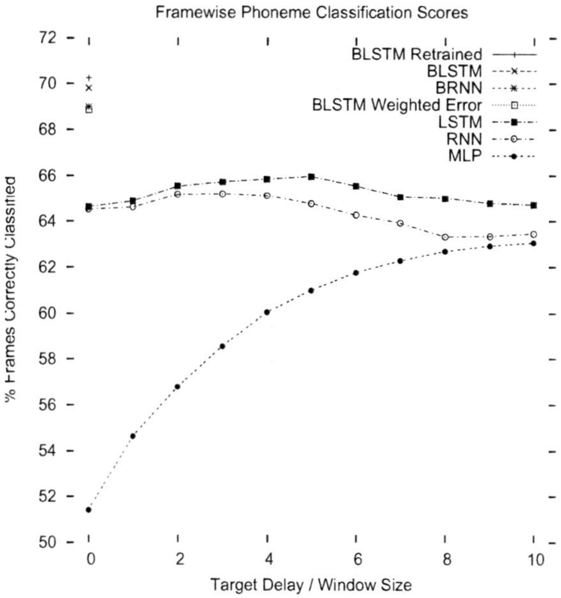


**Fig. 5.2 Framewise phoneme classification results on TIMIT.** The number of frames of added context (time-window size for MLPs, target delay size for unidirectional LSTM and RNNs) is plotted along the x axis. The results for the bidirectional networks (which don't require any extra context) are plotted at x=0.

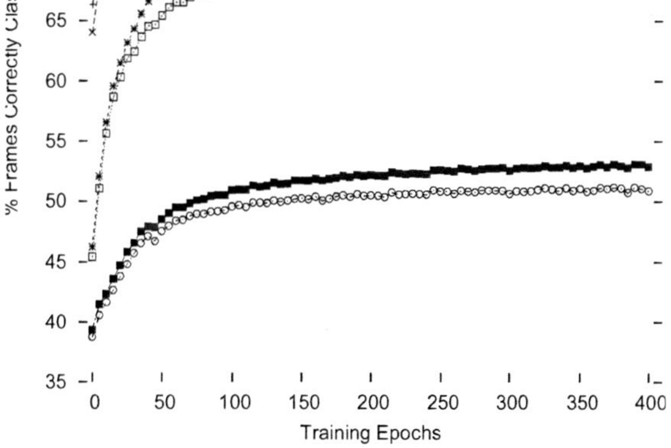


**Fig. 5.3 Learning curves on TIMIT for BLSTM, BRNN and MLP with no time-window.** For all experiments, LSTM was much faster to converge than either the RNN or MLP architectures.

**Table 5.2 Comparison of BLSTM with previous network.** The error measure is the frame error rate (percentage of misclassified frames).

| Network | Train Error (%) | Test Error (%) |
|---|---|---|
| BRNN (Schuster, 1999) | 17.9 | 34.9 |
| RNN (Robinson, 1994) | 29.4 | 34.7 |
| BLSTM (retrained) | 21.4 | 29.8 |
| RNN (Chen and Jamieson, 1996) | 30.1 | 25.8 |

in the way the two architectures learn. Given that in the TIMIT corpus no speakers or sentences are shared by the training and test sets, it is possible that LSTM's overfitting was partly caused by its better adaptation to long range regularities (such as phoneme ordering within words, or speaker specific pronunciations) than normal RNNs. If this is true, we would expect a greater distinction between the two architectures on tasks with more training data.

### *5.4.1 Previous Work*

Table 5.2 shows how BLSTM compares with the best neural network results previously recorded for this task. Note that Robinson did not quote framewise classification scores; the result for his network was recorded by Schuster, using the original software.

Overall BLSTM outperformed all networks found in the literature, apart from the one described by Chen and Jamieson. However this result is questionable as a substantially lower error rate is recorded on the test set than on the training set. Moreover we were unable to reproduce their scores in our own experiments.

In general it is difficult to compare with previous neural network results on this task, owing to variations in network training (different preprocessing, gradient descent algorithms, error functions etc.) and in the task itself (different training and test sets, different numbers of phoneme labels etc.).

### *5.4.2 Effect of Increased Context*

As is clear from Figure 5.2 networks with access to more contextual information tended to get better results. In particular, the bidirectional networks were substantially better than the unidirectional ones. For the unidirectional networks, LSTM benefited more from longer target delays than RNNs; this could be due to LSTM's greater facility with long time-lags, allowing it to make use of the extra context without suffering as much from having to store previous inputs throughout the delay.

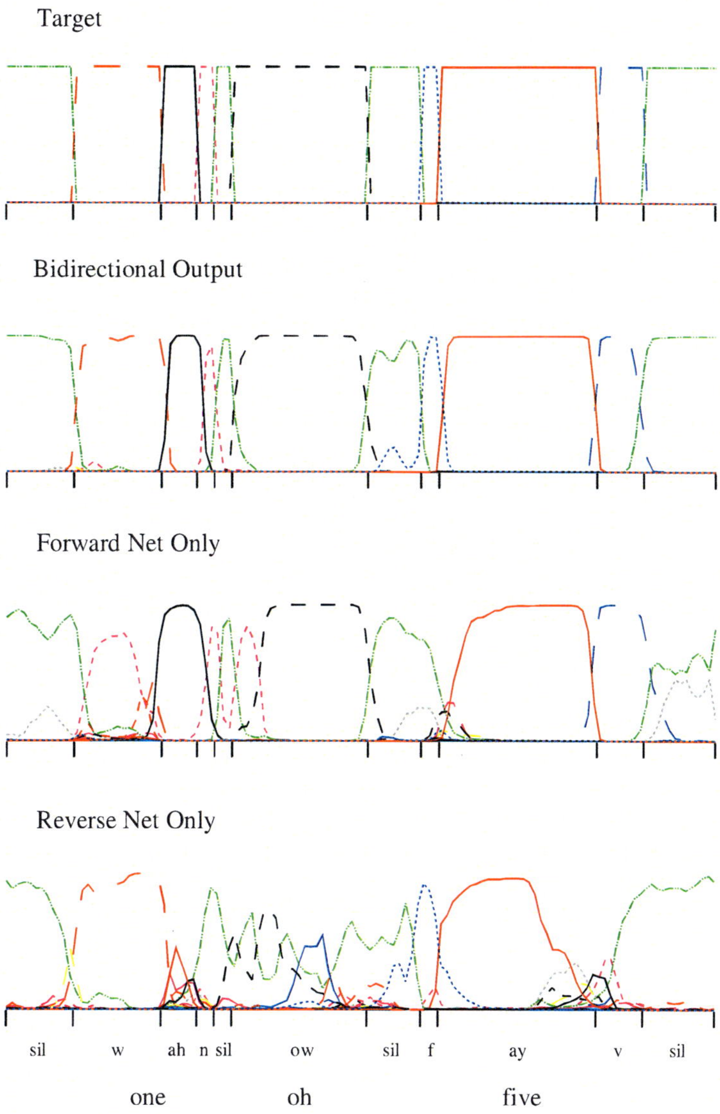


**Fig. 5.4 BLSTM network classifying the utterance "one oh five".** The bidirectional output combines the predictions of the forward and backward hidden layers; it closely matches the target, indicating accurate classification. To see how the layers work together, their contributions to the output are plotted separately. In this case the forward layer seems to be more accurate. However there are places where its substitutions ('w'), insertions (at the start of 'ow') and deletions ('f') appear to be corrected by the backward layer. The outputs for the phoneme 'ay' suggests that the layers can work together, with the backward layer finding the start of the segment and the forward layer finding the end.

Interestingly, LSTM with no time delay returns almost identical results whether trained forwards or backwards. This suggests that the context in both directions is equally important. Figure 5.4 shows how the forward and backward layers work together during classification.

For the MLPs, performance increased with time-window size, and it appears that even larger windows would have been desirable. However, with fully connected networks, the number of weights required for such large input layers makes training prohibitively slow.

### *5.4.3 Weighted Error*

The experiment with a weighted error function gave slightly inferior framewise performance for BLSTM (68.9%, compared to 69.7%). However, the purpose of error weighting is to improve overall phoneme recognition, rather than framewise classification. As a measure of its success, if we assume a perfect knowledge of the test set segmentation (which in real-life situations we cannot), and integrate the network outputs over each phoneme, then BLSTM with weighted errors gives a phoneme error rate of 25.6%, compared to 28.8% with normal errors.

# Chapter 6
# Hidden Markov Model Hybrids

In this chapter LSTM is combined with hidden Markov models (HMMs) to form a hybrid sequence labelling system (Graves et al., 2005b). HMM-neural network hybrids have been extensively studied in the literature, usually with MLPs as the network component. The basic idea is to use the HMM to model the sequential structure of the data, and the neural networks to provide localised classifications. The HMM is able to automatically segment the input sequences during training, and it also provides a principled method for transforming network classifications into label sequences. Unlike the networks described in previous chapters, HMM-ANN hybrids can therefore be directly applied to 'temporal classification' tasks with unsegmented target labels, such as speech recognition.

We evaluate the performance of a hidden Markov model-bidirectional long short-term memory (HMM-BLSTM) hybrid for phoneme recognition, and find that it outperforms both a standard HMM and a hybrid with unidirectional LSTM. This suggests that the advantages of using network architectures with improved contextual processing carry over to temporal classification.

Section 6.1 reviews the previous work on hybrid HMM-neural network systems. Section 6.2 presents experimental results on a phoneme recognition task.

## 6.1 Background

Hybrids of hidden Markov models (HMMs) and artificial neural networks (ANNs) were proposed by several researchers in the 1990s as a way of overcoming the drawbacks of HMMs (Bourlard and Morgan, 1994; Bengio, 1993; Renals et al., 1993; Robinson, 1994; Bengio, 1999). The introduction of ANNs was intended to provide more discriminative training, improved modelling of phoneme duration, richer, nonlinear function approximation, and perhaps most importantly, increased use of contextual information.

In their simplest form, hybrid methods used HMMs to align the segment classifications provided by the ANNs into a temporal classification of the

A. Graves: Supervised Sequence Labell. with Recur. Neur. Networks, SCI 385, pp. 57–60.
springerlink.com 

entire label sequence (Renals et al., 1993; Robinson, 1994). In other cases ANNs were used to estimate transition or emission probabilities for HMMs (Bourlard and Morgan, 1994), to re-score the N-best HMM labellings according to localised classifications (Zavaliagkos et al., 1993), and to extract observation features that can be more easily modelled by an HMM (Bengio et al., 1995, 1992). In this chapter we focus on the simplest case.

Although most hybrid HMM-ANN research has focused on speech recognition, the framework is equally applicable to other sequence labelling tasks, such as online handwriting recognition (Bengio et al., 1995).

The two components in a the hybrid can be trained independently, but many authors have proposed methods for combined optimisation (Bengio et al., 1992; Bourlard et al., 1996; Hennebert et al., 1997; Trentin and Gori, 2003) which typically yields improved results. In this chapter we follow an iterative approach, where the alignment provided by the HMM is used to successively retrain the neural network (Robinson, 1994).

A similar, but more general, framework for combining neural networks with other sequential algorithms is provided by *graph transformer networks* (LeCun et al., 1997, 1998a; Bottou and LeCun, 2005). The different modules of a graph transformer network perform distinct tasks, such as segmentation, recognition and the imposition of grammatical constraints. The modules are connected by *transducers*, which provide differentiable sequence to sequence maps, and allow for global, gradient based learning.

Most hybrid HMM-ANN systems use multilayer perceptrons, typically with a time-window to provide context, for the neural network component. However there has also been considerable interest in the use of RNNs (Robinson, 1994; Neto et al., 1995; Kershaw et al., 1996; Senior and Robinson, 1996). Given that the main purpose of the ANN is to introduce contextual information, RNNs seem a natural choice. However, their advantages over MLPs remained inconclusive in early work (Robinson et al., 1993).

## 6.2 Experiment: Phoneme Recognition

To assess the potential of LSTM and BLSTM for hybrid HMM-ANN systems we compared their performance on the TIMIT speech corpus to that of a standard HMM system. The data preparation and division into training, test and validation sets was identical to that described in Section 5.1. However the task was no longer to phonetically label individual frames, but instead to output the complete phonetic transcription of the sequence. The error measure was therefore the phoneme error rate (label error rate with phonemes as labels—see Section 2.3.3).

We evaluated the performance of standard HMMs with and without context dependent phoneme models, and hybrid systems using BLSTM, LSTM and BRNNs. We also evaluate the effect of using the weighted error signal described in Section 5.2.3.

### *6.2.1 Experimental Setup*

Traditional HMMs were developed with the HTK Speech Recognition Toolkit (http://htk.eng.cam.ac.uk/). Both context independent (mono-phone) and context dependent (triphone) models were trained and tested. Both were left-to-right models with three states. Models representing silence (h#, pau, epi) included two extra transitions, from the first to the final state and vice-versa, to make them more robust. Observation probabilities were modelled by eight Gaussian mixtures.

Sixty-one context-independent models and 5491 tied context-dependent models were used. Context-dependent models for which the left/right context coincide with the central phoneme were included since they appear in the TIMIT transcription (e.g. "my eyes" is transcribed as /m ay ay z/). During recognition, only sequences of context-dependent models with matching context were allowed.

To ensure a fair comparison of the acoustic modelling capabilities of the systems, no prior linguistic information (such as a phonetic language model) was used.

For the hybrid systems, the following networks were used: unidirectional LSTM, BLSTM, and BLSTM trained with weighted error. 61 models of one state each with a self-transition and an exit transition probability were trained using Viterbi-based forced alignment. The frame-level transcription of the training set was used to provide initial estimates of transition and prior probabilities. The networks already trained for the framewise classification experiments in Chapter 5 were re-used for this purpose. The network architectures were therefore identical to those described in Section 5.2.

After initialisation, the hybrid system was trained in an iterative fashion. At each step, the (unnormalised) likelihoods of the 'observations' (i.e. input vectors) conditioned on the hidden states were found by dividing the posterior class probabilities defined by the network outputs by the prior class probabilities found in the data. These likelihoods were used to train the HMM. The alignment provided by the trained HMM was then used to define a new framewise training signal for the neural networks, and the whole process was repeated until convergence. During retraining the network parameters were the same as in Section 5.3, except that Gaussian input noise with a standard deviation of 0.5 was added to the inputs.

For both the standard HMM and the hybrid system, an insertion penalty was optimised on the validation set and applied during recognition.

### *6.2.2 Results*

From Table 6.1, we can see that HMM-BLSTM hybrids outperformed both context-dependent and context-independent HMMs. We can also see that BLSTM gave better performance than unidirectional LSTM, in agreement with the results in Chapter 5. The best result was achieved with the

**Table 6.1 Phoneme recognition results on TIMIT.** The error measure is the phoneme error rate. Hybrid results are means over 5 runs, ± standard error. All differences are significant ($p < 0.01$).

| System | Parameters | Error (%) |
|---|---|---|
| Context-independent HMM | 80 K | 38.85 |
| Context-dependent HMM | >600 K | 35.21 |
| HMM-LSTM | 100 K | $39.6 \pm 0.08$ |
| HMM-BLSTM | 100 K | $33.84 \pm 0.06$ |
| HMM-BLSTM (weighted error) | 100 K | $31.57 \pm 0.06$ |

HMM-BLSTM hybrid using a weighted error signal. This is what we would expect, since the effect of error weighting is to make all phonemes equally significant, as they are to the phoneme error rate.

Note that the hybrid systems had considerably fewer free parameters than the context-dependent HMM. This is a consequence of the high number of states required for HMMs to model contextual dependencies.

The networks in the hybrid systems were initially trained with hand segmented training data. Although the experiments could have been carried out with a flat segmentation, this would probably have led to inferior results.

# Chapter 7
# Connectionist Temporal Classification

This chapter introduces the *connectionist temporal classification* (CTC) output layer for recurrent neural networks (Graves et al., 2006). As its name suggests, CTC was specifically designed for temporal classification tasks; that is, for sequence labelling problems where the alignment between the inputs and the target labels is unknown. Unlike the hybrid approach described in the previous chapter, CTC models all aspects of the sequence with a single neural network, and does not require the network to be combined with a hidden Markov model. It also does not require presegmented training data, or external post-processing to extract the label sequence from the network outputs. Experiments on speech and handwriting recognition show that a BLSTM network with a CTC output layer is an effective sequence labeller, generally outperforming standard HMMs and HMM-neural network hybrids, as well as more recent sequence labelling algorithms such as large margin HMMs (Sha and Saul, 2006) and conditional random fields (Lafferty et al., 2001).

Section 7.1 introduces CTC and motivates its use for temporal classification tasks. Section 7.2 defines the mapping from CTC outputs onto label sequences, Section 7.3 provides an algorithm for efficiently calculating the probability of a given label sequence, Section 7.4 derives the CTC loss function used for network training, Section 7.5 describes methods for decoding with CTC, experimental results are presented in Section 7.6, and a discussion of the differences between CTC networks and HMMs is given in Section 7.7.

## 7.1 Background

In 1994, Bourlard and Morgan identified the following reason for the failure of purely connectionist (that is, neural-network based) approaches to continuous speech recognition:

> There is at least one fundamental difficulty with supervised training of a connectionist network for continuous speech recognition: a target function must be defined, even though the training is

A. Graves: Supervised Sequence Labell. with Recur. Neur. Networks, SCI 385, pp. 61–93.
springerlink.com 

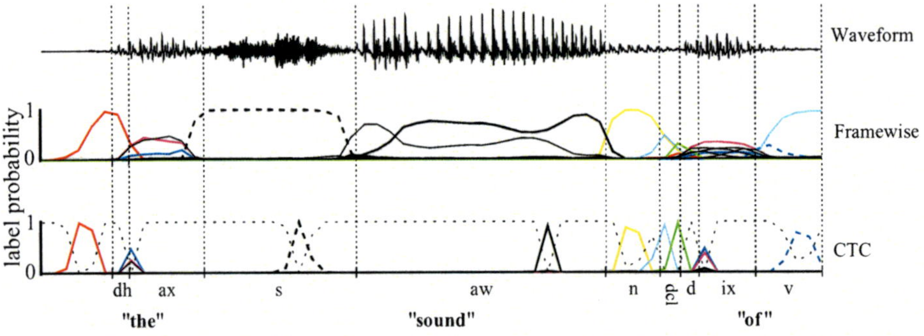


**Fig. 7.1 CTC and framewise classification networks applied to a speech signal.** The coloured lines are the output activations, corresponding to the probabilities of observing phonemes at particular times. The CTC network predicts only the sequence of phonemes (typically as a series of spikes, separated by 'blanks', or null predictions, whose probabilities are shown as a grey dotted line), while the framewise network attempts to align them with the manual segmentation (vertical lines).

> done for connected speech units where the segmentation is generally unknown. (Bourlard and Morgan, 1994, chap. 5)

In other words, neural networks require separate training targets for every segment or timestep in the input sequence. This has two important consequences. Firstly, it means that the training data must be presegmented to provide the targets. Secondly, since the network only outputs local classifications, the global aspects of the sequence, such as the likelihood of two labels appearing consecutively, must be modelled externally. Indeed, without some form of post-processing the final label sequence cannot reliably be inferred at all.

In Chapter 6 we showed how RNNs could be used for temporal classification by combining them with HMMs in hybrid systems. However, as well as inheriting the disadvantages of HMMs (which are discussed in depth in Section 7.7), hybrid systems do not exploit the full potential of RNNs for long-range sequence modelling. It therefore seems preferable to train RNNs directly for temporal classification tasks.

Connectionist temporal classification (CTC) achieves this by allowing the network to make label predictions at any point in the input sequence, so long as the overall sequence of labels is correct. This removes the need for presegmented data, since the alignment of the labels with the input is no longer important. Moreover, CTC directly outputs the probabilities of the complete label sequences, which means that no external post-processing is required to use the network as a temporal classifier.

Figure 7.1 illustrates the difference between CTC and framewise classification applied to a speech signal.

## 7.2 From Outputs to Labellings

For a sequence labelling task where the labels are drawn from an alphabet $A$, CTC consists of a softmax output layer (Bridle, 1990) with one more unit than there are labels in $A$. The activations of the first $|A|$ units are the probabilities of outputting the corresponding labels at particular times, given the input sequence and the network weights. The activation of the extra unit gives the probability of outputting a 'blank', or no label. The complete sequence of network outputs is then used to define a distribution over all possible label sequences of length up to that of the input sequence.

Defining the extended alphabet $A' = A \cup \{blank\}$, the activation $y_k^t$ of network output $k$ at time $t$ is interpreted as the probability that the network will output element $k$ of $A'$ at time $t$, given the length $T$ input sequence $\mathbf{x}$. Let $A'^T$ denote the set of length $T$ sequences over $A'$. Then, if we assume the output probabilities at each timestep to be independent of those at other timesteps (or rather, conditionally independent given $\mathbf{x}$), we get the following conditional distribution over $\pi \in A'^T$:

$$p(\pi|\mathbf{x}) = \prod_{t=1}^{T} y_{\pi_t}^t \tag{7.1}$$

From now on we refer to the sequences $\pi$ over $A'$ as *paths*, to distinguish them from the *label sequences* or *labellings* $\mathbf{l}$ over $A$. The next step is to define a many-to-one function $\mathcal{F} : A'^T \mapsto A^{\leq T}$, from the set of paths onto the set $A^{\leq T}$ of possible labellings of $\mathbf{x}$ (i.e. the set of sequences of length less than or equal to $T$ over $A$). We do this by removing first the repeated labels and then the blanks from the paths. For example $\mathcal{F}(a - ab-) = \mathcal{F}(-aa - -abb) = aab$. Intuitively, this corresponds to outputting a new label when the network either switches from predicting no label to predicting a label, or from predicting one label to another. Since the paths are mutually exclusive, the probability of some labelling $\mathbf{l} \in A^{\leq T}$ can be calculated by summing the probabilities of all the paths mapped onto it by $\mathcal{F}$:

$$p(\mathbf{l}|\mathbf{x}) = \sum_{\pi \in \mathcal{F}^{-1}(\mathbf{l})} p(\pi|\mathbf{x}) \tag{7.2}$$

This 'collapsing together' of different paths onto the same labelling is what makes it possible for CTC to use unsegmented data, because it allows the network to predict the labels without knowing in advance *where* they occur. In theory, it also makes CTC networks unsuitable for tasks where the location of the labels must be determined. However in practice CTC tends to output labels close to where they occur in the input sequence. In Section 7.6.3 an experiment is presented in which both the labels and their approximate positions are successfully predicted by a CTC network.

### *7.2.1 Role of the Blank Labels*

In the original formulation of CTC there were no blank labels, and $\mathcal{F}(\pi)$ was simply $\pi$ with repeated labels removed. This led to two problems. Firstly, the same label could not appear twice in a row, since transitions only occurred when $\pi$ passed between different labels. And secondly, the network was required to continue predicting one label until the next began, which is a burden in tasks where the input segments corresponding to consecutive labels are widely separated by unlabelled data (for example, in speech recognition there are often pauses or non-speech noises between the words in an utterance).

### *7.2.2 Bidirectional and Unidirectional Networks*

Given that the label probabilities used for CTC are assumed to be conditioned on the entire input sequence, it seems natural to prefer a bidirectional RNN architecture. If the network is unidirectional the label probabilities at time $t$ only depend on the inputs up to $t$. The network must therefore wait until after a given input segment is complete (or at least sufficiently complete to be identified) before emitting the corresponding label. This returns us to the issues of past and future context discussed in Chapter 5. Recall that for framewise classification, with a separate target for every input, one way of incorporating future context into unidirectional networks was to introduce a delay between the inputs and the targets. Unidirectional CTC networks are in a somewhat better position, since the delay is not fixed, but can instead be chosen by the network according to the segment being labelled. In practice the performance loss incurred by using unidirectional rather bidirectional RNNs does indeed appear to be smaller for CTC than for framewise classification. This is worth bearing in mind for applications (such as real time speech-recognition) where bidirectional RNNs may be difficult or impossible to apply. Figure 7.2 illustrates some of the differences between unidirectional and bidirectional CTC networks.

## 7.3 Forward-Backward Algorithm

So far we have defined the conditional probabilities $p(\mathbf{l}|\mathbf{x})$ of the possible label sequences. Now we need an efficient way of calculating them. At first sight Eqn. (7.2) suggests this will be problematic: the sum is over all paths corresponding to a given labelling, and the number of these grows exponentially with the length of the input sequence (more precisely, for a length $T$ input sequence and a length $U$ labelling, there are $2^{T-U^2+U(T-3)}3^{(U-1)(T-U)-2}$ paths).

Fortunately the problem can be solved with a dynamic-programming algorithm similar to the forward-backward algorithm for HMMs (Rabiner, 1989).

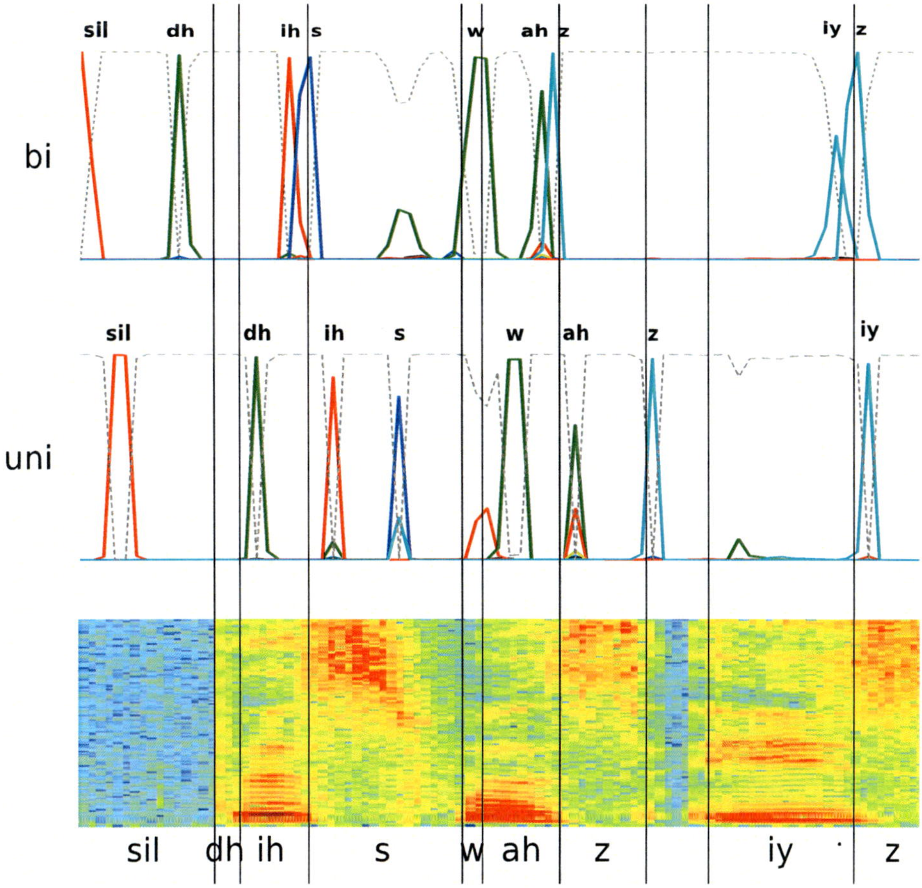


**Fig. 7.2 Unidirectional and Bidirectional CTC Networks Phonetically Transcribing an Excerpt from TIMIT.** The spectrogram (bottom) represents the start of a TIMIT utterance, with the hand segmented phoneme boundaries marked by vertical black lines, and the correct phonetic labels shown underneath. Output phoneme probabilities are indicated by solid coloured lines, while the dashed grey lines correspond to 'blank' probabilities. Both the unidirectional network (middle) and the bidirectional network (top) successfully label the data. However they emit the labels at different times. Whereas the unidirectional network must wait until after the corresponding segments are complete (the exceptions are 'sil' and 's', presumably because they require less context to identify) the bidirectional network may emit the labels before, after or during the segments. Another difference is that bidirectional CTC tends to 'glue together' successive labels that frequently co-occur (for example 'ah' and 'z', which combine to give the rhyming sound in 'was', 'does' and 'buzz').

The key idea is that the sum over paths corresponding to a labelling $\mathbf{l}$ can be broken down into an iterative sum over paths corresponding to prefixes of that labelling.

To allow for blanks in the output paths, we consider a modified label sequence $\mathbf{l}'$, with blanks added to the beginning and the end of $\mathbf{l}$, and inserted between every pair of consecutive labels. If the length of $\mathbf{l}$ is $U$, the length of $\mathbf{l}'$ is therefore $U' = 2U + 1$. In calculating the probabilities of prefixes of $\mathbf{l}'$ we allow all transitions between blank and non-blank labels, and also those between any pair of distinct non-blank labels.

For a labelling $\mathbf{l}$, the *forward variable* $\alpha(t, u)$ is the summed probability of all length $t$ paths that are mapped by $\mathcal{F}$ onto the length $u/2$ prefix of $\mathbf{l}$. For some sequence $\mathbf{s}$, let $\mathbf{s}_{p:q}$ denote the subsequence $\mathbf{s}_p, \mathbf{s}_{p+1}, ..., \mathbf{s}_{q-1}, \mathbf{s}_q$, and let the set $V(t, u) = \{\pi \in A'^t : \mathcal{F}(\pi) = \mathbf{l}_{1:u/2}, \pi_t = l'_u\}$. We can then define $\alpha(t, u)$ as

$$\alpha(t, u) = \sum_{\pi \in V(t,u)} \prod_{i=1}^{t} y^i_{\pi_i} \tag{7.3}$$

where $u/2$ is rounded down to an integer value. As we will see, the forward variables at time $t$ can be calculated recursively from those at time $t - 1$.

Given the above formulation, the probability of $\mathbf{l}$ can be expressed as the sum of the forward variables with and without the final blank at time $T$.

$$p(\mathbf{l}|\mathbf{x}) = \alpha(T, U') + \alpha(T, U' - 1) \tag{7.4}$$

All correct paths must start with either a blank ($b$) or the first symbol in $\mathbf{l}$ ($l_1$), yielding the following initial conditions:

$$\alpha(1, 1) = y^1_b \tag{7.5}$$

$$\alpha(1, 2) = y^1_{l_1} \tag{7.6}$$

$$\alpha(1, u) = 0, \ \forall u > 2 \tag{7.7}$$

Thereafter the variables can be calculated recursively:

$$\alpha(t, u) = y^t_{l'_u} \sum_{i=f(u)}^{u} \alpha(t - 1, i) \tag{7.8}$$

where

$$f(u) = \begin{cases} u - 1 & \text{if } l'_u = \text{blank or } l'_{u-2} = l'_u \\ u - 2 & \text{otherwise} \end{cases} \tag{7.9}$$

Note that

$$\alpha(t, u) = 0 \ \forall u < U' - 2(T - t) - 1 \tag{7.10}$$

because these variables correspond to states for which there are not enough timesteps left to complete the sequence (the unconnected circles in the top right of Figure 7.3). We also impose the boundary condition

$$\alpha(t, 0) = 0 \ \forall t \tag{7.11}$$

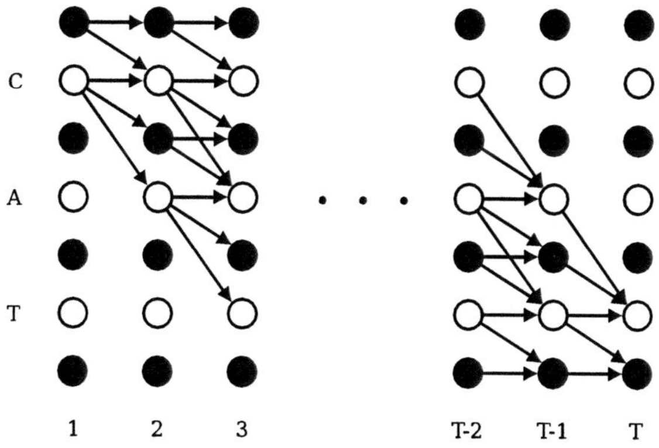


**Fig. 7.3 CTC forward-backward algorithm.** Black circles represent blanks, and white circles represent labels. Arrows signify allowed transitions. Forward variables are updated in the direction of the arrows, and backward variables are updated against them.

The *backward variables* $\beta(t,u)$ are defined as the summed probabilities of all paths starting at $t+1$ that complete $\mathbf{l}$ when appended to any path contributing to $\alpha(t,u)$. Let $W(t,u) = \{\pi \in A'^{T-t} : \mathcal{F}(\hat{\pi} + \pi) = \mathbf{l}\ \forall \hat{\pi} \in V(t,u)\}$. Then

$$\beta(t,u) = \sum_{\pi \in W(t,u)} \prod_{i=1}^{T-t} y^{t+i}_{\pi_i} \tag{7.12}$$

The rules for initialisation and recursion of the backward variables are as follows

$$\beta(T,U') = \beta(T,U'-1) = 1 \tag{7.13}$$

$$\beta(T,u) = 0,\ \forall u < U'-1 \tag{7.14}$$

$$\beta(t,u) = \sum_{i=u}^{g(u)} \beta(t+1,i) y^{t+1}_{l'_i} \tag{7.15}$$

where

$$g(u) = \begin{cases} u+1 & \text{if } l'_u = \text{blank or } l'_{u+2} = l'_u \\ u+2 & \text{otherwise} \end{cases} \tag{7.16}$$

Note that

$$\beta(t,u) = 0\ \forall u > 2t \tag{7.17}$$

as shown by the unconnected circles in the bottom left of Figure 7.3, and

$$\beta(t,U'+1) = 0\ \forall t \tag{7.18}$$

### 7.3.1 *Log Scale*

In practice, the above recursions will soon lead to underflows on any digital computer. A good way to avoid this is to work in the log scale, and only exponentiate to find the true probabilities at the end of the calculation. A useful equation in this context is

$$\ln(a+b) = \ln a + ln\left(1 + e^{\ln b - \ln a}\right) \tag{7.19}$$

which allows the forward and backward variables to be summed while remaining in the log scale. Note that rescaling the variables at every timestep (Rabiner, 1989) is less robust, and can fail for very long sequences.

## 7.4 Loss Function

Like the standard neural network loss functions discussed in Section 3.1.3, the CTC loss function $\mathcal{L}(S)$ is defined as the negative log probability of correctly labelling all the training examples in some training set $S$:

$$\mathcal{L}(S) = -\ln \prod_{(\mathbf{x},\mathbf{z})\in S} p(\mathbf{z}|\mathbf{x}) = -\sum_{(\mathbf{x},\mathbf{z})\in S} \ln p(\mathbf{z}|\mathbf{x}) \tag{7.20}$$

Because the function is differentiable, its derivatives with respect to the network weights can be calculated with backpropagation through time (Section 3.2.2), and the network can then be trained with any gradient-based nonlinear optimisation algorithm (Section 3.3.1).

As in Chapter 2 we also define the *example loss*

$$\mathcal{L}(\mathbf{x},\mathbf{z}) = -\ln p(\mathbf{z}|\mathbf{x}) \tag{7.21}$$

and recall that

$$\mathcal{L}(S) = \sum_{(\mathbf{x},\mathbf{z})\in S} \mathcal{L}(\mathbf{x},\mathbf{z}) \tag{7.22}$$

$$\frac{\partial \mathcal{L}(S)}{\partial w} = \sum_{(\mathbf{x},\mathbf{z})\in S} \frac{\partial \mathcal{L}(\mathbf{x},\mathbf{z})}{\partial w} \tag{7.23}$$

We now show how the algorithm of Section 7.3 can be used to calculate and differentiate $\mathcal{L}(\mathbf{x},\mathbf{z})$, and hence $\mathcal{L}(S)$.

Setting $\mathbf{l} = \mathbf{z}$ and defining the set $X(t,u) = \{\pi \in A'^T : \mathcal{F}(\pi) = \mathbf{z}, \pi_t = \mathbf{z}'_u\}$, Eqns. (7.3) and (7.12) give us

$$\alpha(t,u)\beta(t,u) = \sum_{\pi\in X(t,u)} \prod_{t=1}^{T} y^t_{\pi_t} \tag{7.24}$$

Substituting from (7.1) we get

$$\alpha(t,u)\beta(t,u) = \sum_{\pi\in X(t,u)} p(\pi|\mathbf{x}) \tag{7.25}$$

From (7.2) we can see that this is the portion of the total probability $p(\mathbf{z}|\mathbf{x})$ due to those paths going through $\mathbf{z}'_u$ at time $t$. For any $t$, we can therefore sum over all $u$ to get

$$p(\mathbf{z}|\mathbf{x}) = \sum_{u=1}^{|\mathbf{z}'|} \alpha(t, u)\beta(t, u) \tag{7.26}$$

Meaning that

$$\mathcal{L}(\mathbf{x}, \mathbf{z}) = -\ln \sum_{u=1}^{|\mathbf{z}'|} \alpha(t, u)\beta(t, u) \tag{7.27}$$

### *7.4.1 Loss Gradient*

To find the gradient of $\mathcal{L}(\mathbf{x}, \mathbf{z})$, we first differentiate with respect to the network outputs $y_k^t$:

$$\frac{\partial \mathcal{L}(\mathbf{x}, \mathbf{z})}{\partial y_k^t} = -\frac{\partial \ln p(\mathbf{z}|\mathbf{x})}{\partial y_k^t} = -\frac{1}{p(\mathbf{z}|\mathbf{x})}\frac{\partial p(\mathbf{z}|\mathbf{x})}{\partial y_k^t} \tag{7.28}$$

To differentiate $p(\mathbf{z}|\mathbf{x})$ with respect to $y_k^t$, we need only consider those paths going through label $k$ at time $t$, since the network outputs do not influence each other. Noting that the same label (or blank) may occur several times in a single labelling, we define the set of positions where label $k$ occurs in $\mathbf{z}'$ as $B(\mathbf{z}, k) = \{u : \mathbf{z}'_u = k\}$, which may be empty. Observing from (7.24) that

$$\frac{\partial \alpha(t, u)\beta(t, u)}{\partial y_k^t} = \begin{cases} \frac{\alpha(t, u)\beta(t, u)}{y_k^t} & \text{if } k \text{ occurs in } \mathbf{z}' \\ 0 & \text{otherwise,} \end{cases} \tag{7.29}$$

we can differentiate (7.26) to get

$$\frac{\partial p(\mathbf{z}|\mathbf{x})}{\partial y_k^t} = \frac{1}{y_k^t} \sum_{u \in B(\mathbf{z}, k)} \alpha(t, u)\beta(t, u). \tag{7.30}$$

and substitute this into (7.28) to get

$$\frac{\partial \mathcal{L}(\mathbf{x}, \mathbf{z})}{\partial y_k^t} = -\frac{1}{p(\mathbf{z}|\mathbf{x}) y_k^t} \sum_{u \in B(\mathbf{z}, k)} \alpha(t, u)\beta(t, u). \tag{7.31}$$

Finally, to backpropagate the gradient through the output layer, we need the loss function derivatives with respect to the outputs $a_k^t$ *before* the activation function is applied:

$$\frac{\partial \mathcal{L}(\mathbf{x}, \mathbf{z})}{\partial a_k^t} = -\sum_{k'} \frac{\partial \mathcal{L}(\mathbf{x}, \mathbf{z})}{\partial y_{k'}^t}\frac{\partial y_{k'}^t}{\partial a_k^t} \tag{7.32}$$

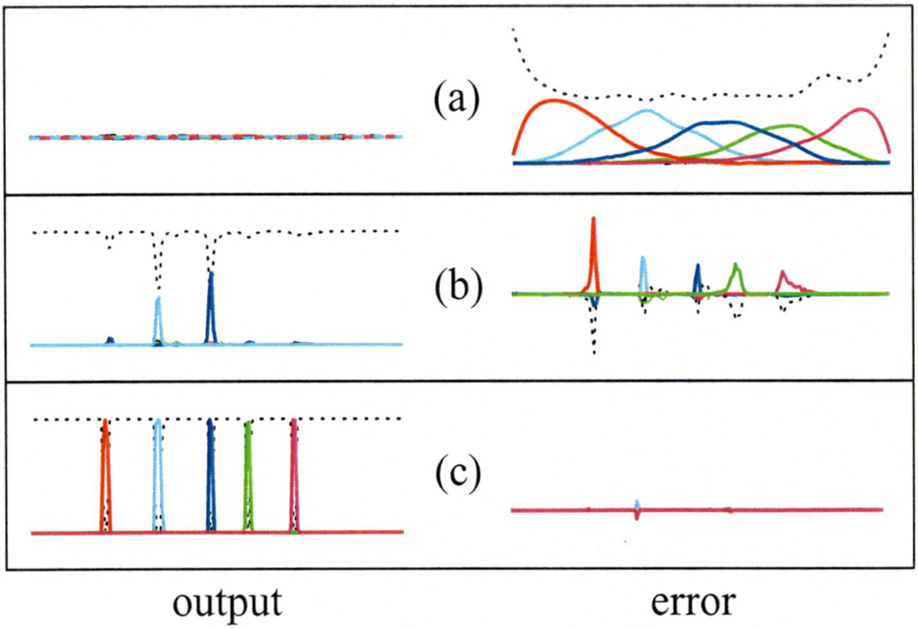


**Fig. 7.4 Evolution of the CTC error signal during training.** The left column shows the output activations for the same sequence at various stages of training (the dashed line is the 'blank' unit); the right column shows the corresponding error signals. Errors above the horizontal axis act to increase the corresponding output activation and those below act to decrease it. (a) Initially the network has small random weights, and the error is determined by the target sequence only. (b) The network begins to make predictions and the error localises around them. (c) The network strongly predicts the correct labelling and the error virtually disappears.

where $k'$ ranges over all the output units. Recalling that for softmax outputs

$$
\begin{aligned}
y^t_k &= \frac{e^{a^t_k}}{\sum_{k'} e^{a^t_{k'}}} \\
\Rightarrow \frac{\partial y^t_{k'}}{\partial a^t_k} &= y^t_{k'}\delta_{kk'} - y^t_{k'}y^t_k
\end{aligned}
\tag{7.33}
$$

we can substitute (7.33) and (7.31) into (7.32) to obtain

$$
\frac{\partial \mathcal{L}(\mathbf{x}, \mathbf{z})}{\partial a^t_k} = y^t_k - \frac{1}{p(\mathbf{z}|\mathbf{x})} \sum_{u \in B(\mathbf{z},k)} \alpha(t,u)\beta(t,u) \tag{7.34}
$$

which is the 'error signal' backpropagated through the network during training, as illustrated in Figure 7.4.

## 7.5 Decoding

Once the network is trained, we would ideally label some unknown input sequence $\mathbf{x}$ by choosing the most probable labelling $\mathbf{l}^*$:

$$\mathbf{l}^* = \arg\max_{\mathbf{l}} p(\mathbf{l}|\mathbf{x}) \tag{7.35}$$

Using the terminology of HMMs, we refer to the task of finding this labelling as *decoding*. Unfortunately, we do not know of a general, tractable decoding algorithm for CTC. However we now present two approximate methods that work well in practice.

### 7.5.1 Best Path Decoding

The first method, which refer to as *best path decoding*, is based on the assumption that the most probable path corresponds to the most probable labelling

$$\mathbf{l}^* \approx \mathcal{F}(\pi^*) \tag{7.36}$$

where $\pi^* = \arg\max_\pi p(\pi|\mathbf{x})$. Best path decoding is trivial to compute, since $\pi^*$ is just the concatenation of the most active outputs at every timestep. However it can lead to errors, particularly if a label is weakly predicted for several consecutive timesteps (see Figure 7.5).

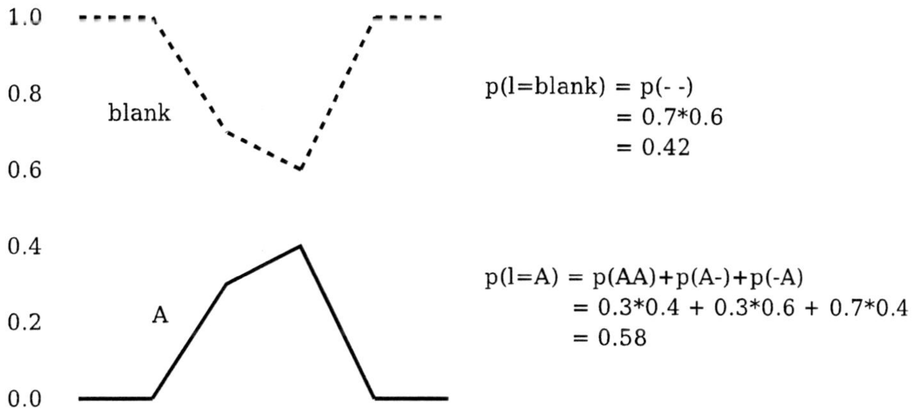


**Fig. 7.5 Problem with best path decoding.** The single most probable path contains no labels, and best path decoding therefore outputs the labelling 'blank'. However the combined probabilities of the paths corresponding to the labelling 'A' is greater.

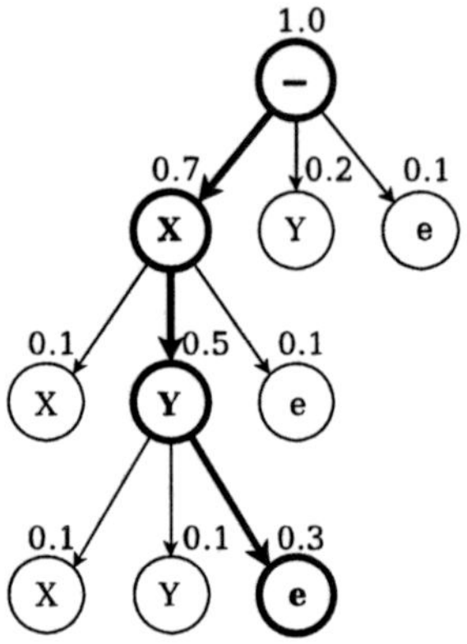


**Fig. 7.6 Prefix search decoding on the alphabet {X,Y}.** Each node either ends ('e') or extends the prefix at its parent node. The number above an extending node is the total probability of all labellings beginning with that prefix. The number above an end node is the probability of the single labelling ending at its parent. At every iteration the extensions of the most probable remaining prefix are explored. Search ends when a single labelling (here 'XY') is more probable than any remaining prefix.

### *7.5.2 Prefix Search Decoding*

The second method (prefix search decoding) relies on the fact that, by modifying the forward variables of Section 7.3, we can efficiently calculate the probabilities of successive extensions of labelling prefixes.

Prefix search decoding is a best-first search (see e.g. Russell and Norvig, 2003, chap. 4) through the tree of labellings, where the children of a given labelling are those that share it as a prefix. At each step the search extends the labelling whose children have the largest cumulative probability (see Figure 7.6).

Let $\gamma(\mathbf{p}_n, t)$ be the probability of the network outputting prefix $\mathbf{p}$ by time $t$ such that a non-blank label is output at $t$, let $\gamma(\mathbf{p}_b, t)$ be the probability of the network outputting prefix $\mathbf{p}$ by time $t$ such that the blank label is output at $t$, and let the set $Y = \{\pi \in A'^t : \mathcal{F}(\pi) = \mathbf{p}\}$. Then

$$\gamma(\mathbf{p}_n, t) = \sum_{\pi \in Y : \pi_t = \mathbf{p}_{|\mathbf{p}|}} p(\pi|\mathbf{x}) \tag{7.37}$$

$$\gamma(\mathbf{p}_b, t) = \sum_{\pi \in Y : \pi_t = blank} p(\pi|\mathbf{x}) \tag{7.38}$$

Thus, for a length $T$ input sequence $\mathbf{x}$, $p(\mathbf{p}|\mathbf{x}) = \gamma(\mathbf{p}_n, T) + \gamma(\mathbf{p}_b, T)$. Also let $p(\mathbf{p}\ldots|\mathbf{x})$ be the cumulative probability of all labellings not equal to $\mathbf{p}$ of which $\mathbf{p}$ is a prefix:

$$p(\mathbf{p}\ldots|\mathbf{x}) = \sum_{\mathbf{l} \neq \emptyset} p(\mathbf{p} + \mathbf{l}|\mathbf{x}) \tag{7.39}$$

where $\emptyset$ is the empty sequence. With these definitions is mind, the pseudocode for prefix search decoding is given in Algorithm 7.1.

Given enough time, prefix search decoding always finds the most probable labelling. However, the maximum number of prefixes it must expand grows exponentially with the input sequence length. If the output distribution is sufficiently peaked around the mode, it will still finish in reasonable time. But for many tasks, a heuristic is required to make its application feasible.

Observing that the outputs of a trained CTC network tend to form a series of spikes separated by strongly predicted blanks (see Figure 7.1), we can divide the output sequence into sections that are very likely to begin and end with a blank. We do this by choosing boundary points where the probability of observing a blank label is above a certain threshold. We then apply Algorithm 7.1 to each section individually and concatenate these to get the final transcription.

In practice, prefix search works well with this heuristic, and generally outperforms best path decoding. However it still makes mistakes in some cases, for example if the same label is predicted weakly on both sides of a section boundary.

### *7.5.3 Constrained Decoding*

For certain tasks we want to constrain the output labellings according to some predefined grammar. For example, in speech and handwriting recognition, the final transcriptions are usually required to form sequences of dictionary words. In addition it is common practice to use a language model to weight the probabilities of particular sequences of words.

We can express these constraints by altering the label sequence probabilities in (7.35) to be conditioned on some probabilistic grammar $G$, as well as the input sequence $\mathbf{x}$

$$\mathbf{l}^* = \arg\max_{\mathbf{l}} p(\mathbf{l}|\mathbf{x}, G) \tag{7.40}$$

Absolute requirements, for example that $\mathbf{l}$ contains only dictionary words, can be incorporated by setting the probability of all sequences that fail to meet them to 0.

At first sight, conditioning on $G$ would seem to contradict a basic assumption of CTC: that the labels are conditionally independent given the input sequences (see Section 7.2). Since the network attempts to model the probability of the whole labelling at once, there is nothing to stop it from learning inter-label transitions direct from the data, which would then be skewed by the external grammar. Indeed, when we tried using a biphone model to decode a CTC network trained for phoneme recognition, the error rate increased. However, CTC networks are typically only able to learn local relationships such as commonly occurring pairs or triples of labels. Therefore as long as $G$ focuses on long range label dependencies (such as the probability of one word following another when the outputs are letters) it doesn't

**Initialisation:**

$1 \leq t \leq T \begin{cases} \gamma(\emptyset_n, t) = 0 \\ \gamma(\emptyset_b, t) = \prod_{t'=1}^{t} y_b^{t'} \end{cases}$

$p(\emptyset|\mathbf{x}) = \gamma(\emptyset_b, T)$

$p(\emptyset \ldots |\mathbf{x}) = 1 - p(\emptyset|\mathbf{x})$

$\mathbf{l}^* = \mathbf{p}^* = \emptyset$

$P = \{\emptyset\}$

**Algorithm:**

**while** $p(\mathbf{p}^* \ldots |\mathbf{x}) > p(\mathbf{l}^*|\mathbf{x})$ **do**

  $probRemaining = p(\mathbf{p}^* \ldots |\mathbf{x})$

  **for** all labels $k \in A$ **do**

    $\mathbf{p} = \mathbf{p}^* + k$

    $\gamma(\mathbf{p}_n, 1) = \begin{cases} y_k^1 \text{ if } \mathbf{p}^* = \emptyset \\ 0 \text{ otherwise} \end{cases}$

    $\gamma(\mathbf{p}_b, 1) = 0$

    $prefixProb = \gamma(\mathbf{p}_n, 1)$

    **for** $t = 2$ **to** $T$ **do**

      $newLabelProb = \gamma(\mathbf{p}_b^*, t-1) + \begin{cases} 0 \text{ if } \mathbf{p}^* \text{ ends in } k \\ \gamma(\mathbf{p}_n^*, t-1) \text{ otherwise} \end{cases}$

      $\gamma(\mathbf{p}_n, t) = y_k^t \left(newLabelProb + \gamma(\mathbf{p}_n, t-1)\right)$

      $\gamma(\mathbf{p}_b, t) = y_b^t \left(\gamma(\mathbf{p}_b, t-1) + \gamma(\mathbf{p}_n, t-1)\right)$

      $prefixProb \mathrel{+}= y_k^t \; newLabelProb$

    $p(\mathbf{p}|\mathbf{x}) = \gamma(\mathbf{p}_n, T) + \gamma(\mathbf{p}_b, T)$

    $p(\mathbf{p} \ldots |\mathbf{x}) = prefixProb - p(\mathbf{p}|\mathbf{x})$

    $probRemaining \mathrel{-}= p(\mathbf{p} \ldots |\mathbf{x})$

    **if** $p(\mathbf{p}|\mathbf{x}) > p(\mathbf{l}^*|\mathbf{x})$ **then**

      $\mathbf{l}^* = \mathbf{p}$

    **if** $p(\mathbf{p} \ldots |\mathbf{x}) > p(\mathbf{l}^*|\mathbf{x})$ **then**

      add $\mathbf{p}$ to $P$

    **if** $probRemaining \leq p(\mathbf{l}^*|\mathbf{x})$ **then**

      break

  remove $\mathbf{p}^*$ from $P$

  $\mathbf{p}^* = \arg\max_{\mathbf{p} \in P} \; p(\mathbf{p} \ldots |\mathbf{x})$

**Termination:**

output $\mathbf{l}^*$

**Algorithm 7.1** Prefix Search Decoding Algorithm

interfere with the dependencies modelled internally by CTC. This argument is supported by the experiments in Sections 7.6.4 and 7.6.5.

Applying the basic rules of probability we obtain

$$p(\mathbf{l}|\mathbf{x}, G) = \frac{p(\mathbf{l}|\mathbf{x})p(\mathbf{l}|G)p(\mathbf{x})}{p(\mathbf{x}|G)p(\mathbf{l})} \tag{7.41}$$

where we have used the fact that $\mathbf{x}$ is conditionally independent of $G$ given $\mathbf{l}$. If we assume that $\mathbf{x}$ is independent of $G$, (7.41) reduces to

$$p(\mathbf{l}|\mathbf{x}, G) = \frac{p(\mathbf{l}|\mathbf{x})p(\mathbf{l}|G)}{p(\mathbf{l})} \tag{7.42}$$

This assumption is in general false, since both the input sequences and the grammar depend on the underlying generator of the data, for example the language being spoken. However it is a reasonable first approximation, and is particularly justifiable in cases where the grammar is created using data other than that from which $\mathbf{x}$ was drawn (as is common practice in speech and handwriting recognition, where separate textual corpora are used to generate language models).

If we further assume that, prior to any knowledge about the input or the grammar, all label sequences are equally probable, (7.40) reduces to

$$\mathbf{l}^* = \arg\max_{\mathbf{l}} p(\mathbf{l}|\mathbf{x})p(\mathbf{l}|G) \tag{7.43}$$

Note that, since the number of possible label sequences is finite (because both $A$ and $S$ are finite), assigning equal prior probabilities does not lead to an improper prior.

#### 7.5.3.1 CTC Token Passing Algorithm

We now describe an algorithm, based on the *token passing algorithm* for HMMs (Young et al., 1989), that finds an approximate solution to (7.43) for a simple grammar.

Let $G$ consist of a dictionary $D$ containing $W$ words, and an optional set of $W^2$ bigrams $p(w|\hat{w})$ that define the probability of making a transition from word $\hat{w}$ to word $w$. The probability of any label sequence that does not form a sequence of dictionary words is 0.

For each word $w$, define the modified word $w'$ as $w$ with blanks added at the beginning and end and between each pair of labels. Therefore $|w'| = 2|w|+1$. Define a token $tok = (score, history)$ to be a pair consisting of a real valued 'score' and a 'history' of previously visited words. The history corresponds to the path through the network outputs the token has taken so far, and the score is the log probability of that path. The basic idea of the token passing algorithm is to pass along the highest scoring tokens at every word state, then maximise over these to find the highest scoring tokens at the next state. The transition probabilities are used when a token is passed from the last state in one word to the first state in another. The output word sequence is then given by the history of the highest scoring end-of-word token at the final timestep.

At every timestep $t$ of the length $T$ output sequence, each segment $s$ of each modified word $w'$ holds a single token $tok(w, s, t)$. This is the highest scoring token reaching that segment at that time. Define the *input token* $tok(w, 0, t)$ to be the highest scoring token arriving at word $w$ at time $t$, and the *output token* $tok(w, -1, t)$ to be the highest scoring token leaving word $w$ at time $t$. $\emptyset$ denotes the empty sequence.

Pseudocode for the algorithm is provided in Algorithm 7.2.

#### 7.5.3.2 Computational Complexity

If bigrams are used, the CTC token passing algorithm has a worst-case complexity of $\mathcal{O}(TW^2)$, since line 19 requires a potential search through all $W$ words. However, because the output tokens $tok(w, -1, T)$ are sorted in order of score, the search can be terminated when a token is reached whose score is less than the current best score with the transition included. The typical complexity is therefore considerably lower, with a lower bound of $\mathcal{O}(TWlogW)$ to account for the sort.

If no bigrams are used, the single most probable output token at the previous timestep will form the new input token for all the words, and the worst-case complexity reduces to $\mathcal{O}(TW)$.

#### 7.5.3.3 Single Word Decoding

If the number of words in the target sequence is fixed, Algorithm 7.2 can be constrained by forbidding all tokens whose history already contains that many words from transitioning to new words. In particular, if the target sequences are constrained to be single words, then all word-to-word transitions are forbidden (and bigrams are clearly not required).

In general the extension from finding the single best transcription to the $N$-best transcriptions is complex. However in the special case of single word decoding, the $N$-best transcriptions are simply the (single word) histories of the $N$-best output tokens when the algorithm terminates.

Another straightforward extension to single word decoding occurs when the same word has several different label transcriptions. This happens, for example, when pronunciation variants are considered in speech recognition, or spelling variants are allowed in handwriting recognition. In that case all variants should be considered separate words until the termination of Algorithm 7.2 (lines 34 and 34); at that point the scores of all variant transcriptions of each word should be added together *in the log scale* (that is, using Eqn. (7.19)); thereafter the best or $N$-best words should be found as usual.

Several tasks requiring $N$-best single word transcriptions, both with and without transcription variants, are presented in Chapter 9.

```
Initialisation:
for all words w ∈ D do
    tok(w, 1, 1) = (ln y^1_b, (w))
    tok(w, 2, 1) = (ln y^1_{w_1}, (w))
    if |w| = 1 then
        tok(w, −1, 1) = tok(w, 2, 1)
    else
        tok(w, −1, 1) = (−∞, ∅)
    tok(w, s, 1) = (−∞, ∅) for all other s

Algorithm:
for t = 2 to T do
    if using bigrams then
        sort output tokens tok(w, −1, t − 1) by ascending score
    else
        find single highest scoring output token
    for all words w ∈ D do
        if using bigrams then
            w* = arg max_ŵ [tok(ŵ, −1, t − 1).score + ln p(w|ŵ)]
            tok(w, 0, t) = tok(w*, −1, t − 1)
            tok(w, 0, t).score += ln p(w|w*)
        else
            tok(w, 0, t) = highest scoring output token
        add w to tok(w, 0, t).history
        for segment s = 1 to |w'| do
            P = {tok(w, s, t − 1), tok(w, s − 1, t − 1)}
            if w'_s ≠ blank and s > 2 and w'_{s−2} ≠ w'_s then
                add tok(w, s − 2, t − 1) to P
            tok(w, s, t) = token in P with highest score
            tok(w, s, t).score += ln y^t_{w'_s}
        tok(w, −1, t) = highest scoring of {tok(w, |w'|, t), tok(w, |w'| − 1, t)}

Termination:
w* = arg max_w tok(w, −1, T).score
output tok(w*, −1, T).history
```

**Algorithm 7.2** CTC Token Passing Algorithm

## 7.6 Experiments

In this section bidirectional LSTM (BLSTM) networks with CTC output layers are evaluated on five temporal classification tasks: three related to speech recognition, and two related to handwriting recognition.

For the handwriting tasks, a dictionary and language model were present, and results are recorded both with and without the constrained decoding algorithm of Section 7.5.3.1. For the speech experiments there was no dictionary or language model, and the output labels (whether phonemes or whole words) were used directly for transcription. For the experiment in Section 7.6.1, we compare prefix search and best path decoding (see Section 7.5).

As discussed in Chapter 3, the choice of input representation is crucial to any machine learning algorithm. Most of the experiments here use standard input representations that have been tried and tested with other sequence learning algorithms, such as HMMs. The experiment in Section 7.6.4 is different in that the performance of BLSTM-CTC is compared using two different input representations. As usual, all inputs were standardised to have mean 0 and standard deviation 1 over the training set.

For all the experiments, the BLSTM hidden layers were fully connected to themselves, and to the input and output layers. Each memory block contained a single LSTM cell, with tanh used for the activation functions $g$ and $h$ and the logistic sigmoid $\sigma(x) = 1/(1 + e^{-x})$ used for the activation function $f$ of the gates (see Figure 4.2). The sizes of the input and output layers were determined by the numbers of inputs and labels in each task. The weights were randomly initialised from a Gaussian distribution with mean 0 and standard deviation 0.1. Online steepest descent with momentum was used for training, with a learning rate $10^{-4}$ and a momentum of 0.9. All experiments used separate training, validation and testing sets. Training was stopped when 50 epochs had passed with no reduction of error on the validation set.

The only network parameters manually adjusted for the different tasks were (1) the number of blocks in the LSTM layers and (2) the standard deviation of the input noise added during training. These are specified for each experiment.

As discussed in Section 3.3.2.2, Gaussian input noise is only effective if it mimics the true variations found in the data. This appears to be the case for the MFC coefficients commonly used for speech recognition, but not for the other data types considered below. Therefore input noise was only applied for the speech recognition experiments.

For all experiments, the error measure used to evaluate performance was the *label error rate* defined in Section 2.3.3, applied to the test set. Note however that the measure is renamed according to the type of labels used: for example *phoneme error rate* was used for phoneme recognition, and *word error rate* for keyword spotting. For the handwriting recognition tasks, both the *character error rate* for the labellings provided by the CTC output layer, and the *word error rate* for the word sequences obtained from the token

**Table 7.1 Phoneme recognition results on TIMIT with 61 phonemes.** The error measure is the phoneme error rate. BLSTM-CTC and hybrid results are means over 5 runs, ± standard error. All differences were significant ($p < 0.01$), except that between HMM-BLSTM with weighted errors and CTC with best path decoding.

| System | Error (%) |
|---|---|
| HMM | 35.21 |
| HMM-BLSTM hybrid (weighted error) | 31.57 ± 0.06 |
| BLSTM-CTC (best path decoding) | 31.47 ± 0.21 |
| BLSTM-CTC (prefix search decoding) | 30.51 ± 0.19 |

passing algorithm are evaluated. All the experiments were repeated several times and the results are quoted as a mean ± the standard error. The mean and standard error in the number of training epochs before the best results were achieved is also provided.

## *7.6.1 Phoneme Recognition 1*

The experiment in this section compares BLSTM-CTC with the best HMM and HMM-BLSTM hybrid results given in Chapter 6 for phoneme recognition on the TIMIT speech corpus (Garofolo et al., 1993). The task, data and preprocessing were identical to those described in Section 5.1.

### 7.6.1.1 Experimental Setup

The network had 26 input units and 100 LSTM blocks in both the forward and backward hidden layers. It had 62 output units, one for each phoneme plus the 'blank', giving 114,662 weights in total. Gaussian noise with mean 0 and standard deviation 0.6 was added to the inputs during training. When prefix search decoding was used, the probability threshold for the boundary points was 0.9999.

### 7.6.1.2 Results

Table 7.1 shows that, with prefix search decoding, BLSTM-CTC outperformed both an HMM and an HMM-RNN hybrid with the same RNN architecture. It also shows that prefix search gave a small improvement over best path decoding.

Note that the best hybrid results were achieved with a weighted error signal. Such heuristics are unnecessary with CTC, as the loss function depends only on the *sequence* of labels, and not on their duration or segmentation.

Input noise had a greater impact on generalisation for BLSTM-CTC than the hybrid system, and a slightly higher level of noise was found to be optimal

**Table 7.2 Folding the 61 phonemes in TIMIT onto 39 categories** (Lee and Hon, 1989). The phonemes in the right column are folded onto the corresponding category in the left column ('q' is discarded). All other phonemes are left unchanged.

| | |
|---|---|
| aa | aa, ao |
| ah | ah, ax, ax-h |
| er | er, axr |
| hh | hh, hv |
| ih | ih, ix |
| l | l, el |
| m | m, em |
| n | n, en, nx |
| ng | ng, eng |
| sh | sh, zh |
| sil | pcl, tcl, kcl, bcl, dcl, gcl, h#, pau, epi |
| uw | uw, ux |
| — | q |

(standard deviation 0.6 for the CTC network and 0.5 for the hybrid). The mean training time for BLSTM-CTC was 60.0 $\pm$ 7 epochs.

### *7.6.2 Phoneme Recognition 2*

This section considers a variant of the previous task, where the number of distinct phoneme labels is reduced from 61 to 39 (Fernndez et al., 2008). In addition, only the so-called *core test set* of TIMIT is used for evaluation. These modifications allow a direct comparison with other results in the literature.

#### 7.6.2.1 Data and Preprocessing

In most previous studies, a set of 48 phonemes were selected for modelling during training, and confusions between several of these were ignored during testing, leaving an effective set of 39 distinct labels (Lee and Hon, 1989). Since CTC is discriminative, using extra phonemes during training is unnecessary (and probably counterproductive), and the networks were therefore trained with 39 labels. The folding of the original 61 phoneme labels onto 39 categories is shown in table 7.2.

The TIMIT corpus was divided into a training set, a validation set and a test set according to (Halberstadt, 1998). As in our previous experiments, the training set contained 3696 sentences from 462 speakers. However in this case the test set was much smaller, containing only 192 sentences from 24 speakers, and the validation set, which contained 400 sentences from 50 speakers, was drawn from the unused test sentences rather than the training set. This left us with slightly more sentences for training than before. However this advantage was offset by the fact that the core test set is harder than the full test set.

As before, the speech data was transformed into Mel frequency cepstral coefficients (MFCCs) using the HTK software package (Young et al., 2006). Spectral analysis was carried out with a 40 channel Mel filter bank from 64 Hz to 8 kHz. A pre-emphasis coefficient of 0.97 was used to correct spectral tilt. Twelve MFCCs plus the 0th order coefficient were computed on Hamming windows 25 ms long, every 10 ms. In this case the second as well as first derivatives of the coefficients were used, giving a vector of 39 inputs in total.

#### 7.6.2.2 Experimental Setup

The BLSTM-CTC network had an input layer of size 39, the forward and backward hidden layers had 128 blocks each, and the output layer was size 40 (39 phonemes plus blank). The total number of weights was 183,080. Gaussian noise with a standard deviation of 0.6 was added to the inputs during training. When prefix search was used, the probability threshold was 0.9999.

#### 7.6.2.3 Results

The performance of the network is recorded in table 7.3, along with the best results found in the literature. There is no significant difference between the error rate of BLSTM-CTC with prefix search decoding and that of At the time when the experiment was performed there was no significant difference between the BLSTM-CTC results and that of either of the best two results found in the literature (by Yu *et al* and Glass). However the benchmark has since been substantially lowered by the application of improved language modelling techniques (Sainath et al., 2009) and Deep Belief Networks (Mohamed et al., 2011).

Nonetheless the BLSTM-CTC score remains a good result for a general purpose sequence labelling system with very little tuning towards speech recognition.

### *7.6.3 Keyword Spotting*

The task in this section is keyword spotting, using the Verbmobil speech corpus (Verbmobil, 2004). The aim of keyword spotting is to identify a particular set of spoken words within (typically unconstrained) speech signals. In most cases, the keywords occupy only a small fraction of the total data. Discriminative approaches are interesting for keyword spotting, because they are able to concentrate on identifying and distinguishing the keywords, while ignoring the rest of the signal. However, the predominant method is to use hidden Markov models, which are generative, and must therefore model the unwanted speech, and even the non-speech noises, in addition to the keywords.

**Table 7.3 Phoneme recognition results on TIMIT with 39 phonemes.** The error measure is the phoneme error rate. Results for BLSTM-CTC are averages ± standard error over 10 runs. The average number of training epochs was 112.5 ± 6.4.

| System | Error (%) |
|---|---|
| Conditional Random Fields (Morris and Lussier, 2006) | 34.8 |
| Large Margin HMM (Sha and Saul, 2006) | 28.7 |
| Baseline HMM (Yu et al., 2006) | 28.6 |
| Triphone Continuous Density HMM (Lamel and Gauvain, 1993) | 27.1 |
| Augmented Conditional Random Fields (Hifny and Renals, 2009) | 26.7 |
| RNN-HMM Hybrid (Robinson, 1994) | 26.1 |
| Bayesian Triphone HMM (Ming and Smith, 1998) | 25.6 |
| Near-miss Probabilistic Segmentation (Chang, 1998) | 25.5 |
| BLSTM-CTC (best path decoding) | 25.2 ± 0.2 |
| Monophone HTMs (Yu et al., 2006) | 24.8 |
| BLSTM-CTC (prefix search decoding) | 24.6 ± 0.2 |
| Heterogeneous Classifiers (Glass, 2003) | 24.4 |
| Discriminative BMMI Triphone HMMs (Sainath et al., 2009) | 22.7 |
| Monophone Deep Belief Network (Mohamed et al., 2011) | 20.7 |

In many cases one seeks not only the identity of the keywords, but also their approximate position. For example, this would be desirable if the goal were to further examine those segments of a long telephone conversation in which a keyword occurred. In principle, locating the keywords presents a problem for CTC, since the network is only trained to find the sequence of labels, and not their position. However we have observed that in most cases CTC predicts labels close to the relevant segments of the input sequence. The following experiments confirm this observation by recording the word error rate both with and without the requirement that the network find the approximate location of the keywords (Fernández et al., 2007).

#### 7.6.3.1 Data and Preprocessing

Verbmobil consists of dialogues of noisy, spontaneous German speech, where the purpose of each dialogue is to schedule a date for an appointment or meeting. It comes divided into training, validation and testing sets, all of which have a complete phonetic transcription. The training set includes 748 speakers and 23,975 dialogue turns, giving a total of 45.6 hours of speech. The validation set includes 48 speakers, 1,222 dialogue turns and a total of 2.9 hours of speech. The test set includes 46 speakers, 1,223 dialogue turns and a total of 2.5 hours of speech. Each speaker appears in only one of the sets.

The twelve keywords were: *April, August, Donnerstag, Februar, Frankfurt, Freitag, Hannover, Januar, Juli, Juni, Mittwoch, Montag.* Since the

**Table 7.4 Keyword spotting results on Verbmobil.** The error measure is the keyword error rate. Results are a mean over 4 runs, ± standard error.

| System | Error (%) |
|---|---|
| BLSTM-CTC (approx. location) | $15.5 \pm 1.2$ |
| BLSTM-CTC (any location) | $13.9 \pm 0.7$ |

dialogues are concerned with dates and places, all of these occurred fairly frequently in the data sets. One complication is that there are pronunciation variants of some of these keywords (e.g. "Montag" can end either with a /g/ or with a /k/). Another is that several keywords appear as sub-words, e.g. in plural form such as "Montags" or as part of another word such as "Ostermontag" (Easter Monday). The start and end times of the keywords were given by the automatic segmentation provided with the phonetic transcription.

In total there were 10,469 keywords on the training set with an average of 1.7% keywords per non-empty utterance (73.6% of the utterances did not have any keyword); 663 keywords on the validation set with an average of 1.7% keywords per non-empty utterance (68.7% of the utterances did not have any keyword); and 620 keywords on the test set with an average of 1.8 keywords per non-empty utterance (71.1% of the utterances did not have any keyword).

The audio preprocessing was identical to that described in Section 5.1, except that the second order derivatives of the MFCC coefficients were also included, giving a total of 39 inputs per frame.

#### 7.6.3.2 Experimental Setup

The BLSTM-CTC network contained 128 single-memory-cell LSTM blocks in the forward and backward hidden layers. The output layer contained 13 units and the input layer contained 39 units, giving 176,141 weights in total. Gaussian noise with a mean of 0 and a standard deviation of 0.5 was added during training.

We considered the network to have successfully found the approximate location of a keyword if it predicted the correct label within 0.5 seconds of the boundary of the keyword segment. The experiment was not repeated for the approximate location results: the output of the network was simply re-scored with the location constraint included.

#### 7.6.3.3 Results

Table 7.4 shows that the network gave a mean error rate of 15.5%. Performance was only slightly better without the constraint that it find the approximate location of the keywords. This shows in most cases it aligned the keywords with the relevant portion of the input signal.

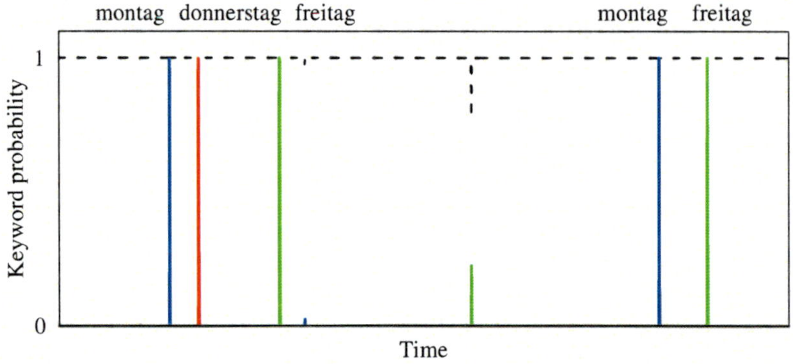


**Fig. 7.7 CTC outputs for keyword spotting on Verbmobil**

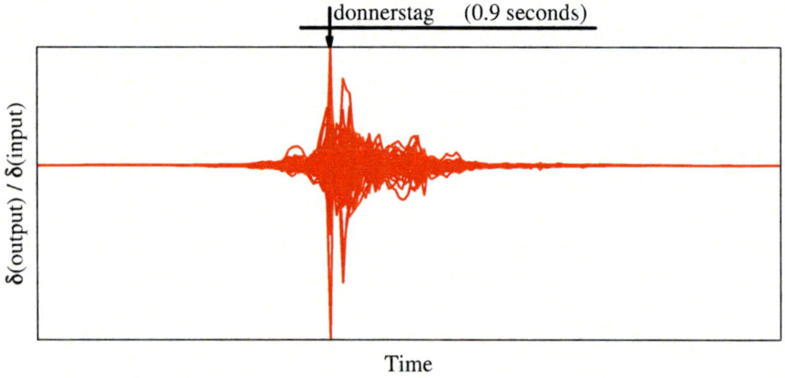


**Fig. 7.8 Sequential Jacobian for keyword spotting on Verbmobil**

Although we don't have a direct comparison for this result, a benchmark HMM system performing full speech recognition on the same dataset achieved a word error rate of 35%. We attempted to train an HMM system specifically for keyword spotting, with a single junk model for everything apart from the keywords, but found that it did not converge. This is symptomatic of the difficulty of using a generative model for a task where so much of the input is irrelevant.

The mean training time for the network was $91.3 \pm 22.5$ epochs.

#### 7.6.3.4 Analysis

Figure 7.7 shows the CTC outputs during a dialogue turn containing several keywords. For a zoomed in section of the same dialogue turn, Figure 7.8 shows the sequential Jacobian for the output unit associated with the keyword "Donnerstag" at the time step indicated by an arrow at the top of the figure. The extent (0.9 s) and location of the keyword in the speech signal is shown at the top of the figure. As can be seen, the output is most sensitive to the first part of the keyword. This is unsurprising, since the ending, "tag", is shared by many of the keywords and is therefore the least discriminative part.

## 7.6.4 Online Handwriting Recognition

The task in this section is online handwriting recognition, using the IAM-OnDB handwriting database (Liwicki and Bunke, 2005b)[1]. In online handwriting recognition, the state and position of the pen is recorded during writing, and used as input to the learning algorithm.

For the CTC experiments (Liwicki et al., 2007; Graves et al., 2008, 2009), the character error rate is obtaining using best-path decoding, and the word error rate using constrained decoding. For the HMM system, only the word error rate is given.

We compare results using two different input representations, one hand crafted for HMMs, the other consisting of raw data direct from the pen sensor.

### 7.6.4.1 Data and Preprocessing

IAM-OnDB consists of pen trajectories collected from 221 different writers using a 'smart whiteboard' (Liwicki and Bunke, 2005a). The writers were asked to write forms from the LOB text corpus (Johansson et al., 1986), and the position of their pen was tracked using an infra-red device in the corner of the board. The original input data consists of the $x$ and $y$ pen coordinates, the points in the sequence when individual strokes (i.e. periods when the pen is pressed against the board) end, and the times when successive position measurements were made. Recording errors in the $x, y$ data was corrected by interpolating to fill in for missing readings, and removing steps whose length exceeded a certain threshold.

The character level transcriptions contain 80 distinct target labels (capital letters, lower case letters, numbers, and punctuation). A dictionary consisting of the 20,000 most frequently occurring words in the LOB corpus was used for decoding, along with a bigram language model. 5.6% of the words in the test set were not contained in the dictionary. The language model was optimised on the training and validation sets only.

IAM-OnDB is divided into a training set, two validation sets, and a test set, containing respectively 5364, 1,438, 1,518 and 3,859 written lines taken from 775, 192, 216 and 544 forms. For our experiments, each line was assumed to be an independent sequence (meaning that the dependencies between successive lines, e.g. for a continued sentence, were ignored).

Two input representations were used for this task. The first consisted simply of the offset of the $x, y$ coordinates from the top left of the line, along with the time from the beginning of the line, and an extra input to mark the

[1] Available for public download at http://www.iam.unibe.ch/~fki/iamondb/

**Table 7.5 Character recognition results on IAM-OnDB.** The error measure is the character error rate. Results are a mean over 4 runs, ± standard error.

| Input | Error (%) |
|---|---|
| Raw | 13.9 ± 0.1 |
| Preprocessed | 11.5 ± 0.05 |

points when the pen was lifted off the whiteboard (see Figure 7.9). We refer to this as the *raw* representation.

The second representation required a large amount of sophisticated preprocessing and feature extraction (Liwicki et al., 2007). We refer to this as the *preprocessed* representation. Briefly, in order to account for the variance in writing styles, the pen trajectories were first normalised with respect to such properties as the slant, skew and width of the letters, and the slope of the line as a whole. Two sets of input features were then extracted, one consisting of 'online' features, such as pen position, pen speed, line curvature etc., and the other consisting of 'offline' features derived from a two dimensional window of the image reconstructed from the pen trajectory. Delayed strokes (such as the crossing of a 't' or the dot of an 'i') are removed by the preprocessing because they introduce difficult long time dependencies.

#### 7.6.4.2 Experimental Setup

The network contained 100 LSTM blocks in the forward and backward hidden layers. The output layer contained 81 units. For the raw representation, there were 4 input units, giving 100,881 weights in total. For the preprocessed representation, there were 25 input units, giving 117,681 weights in total. No noise was added during training.

The HMM setup (Liwicki et al., 2007) contained a separate, linear HMM with 8 states for each character ($8 * 81 = 648$ states in total). Diagonal mixtures of 32 Gaussians were used to estimate the observation probabilities. All parameters, including the word insertion penalty and the grammar scale factor, were optimised on the validation set.

#### 7.6.4.3 Results

Table 7.6 shows that with a language model and the preprocessed input representation, BLSTM-CTC gives a mean word error rate of 20.4%, compared to 35.5% with a benchmark HMM. This is an error reduction of 42.5%. Moreover, even without the language model or the handcrafted preprocessing, BLSTM-CTC clearly outperforms HMMs.

The mean training time for the network was 41.3 ± 2.4 epochs for the preprocessed data, and 233.8 ± 16.8 epochs for the raw data. This disparity

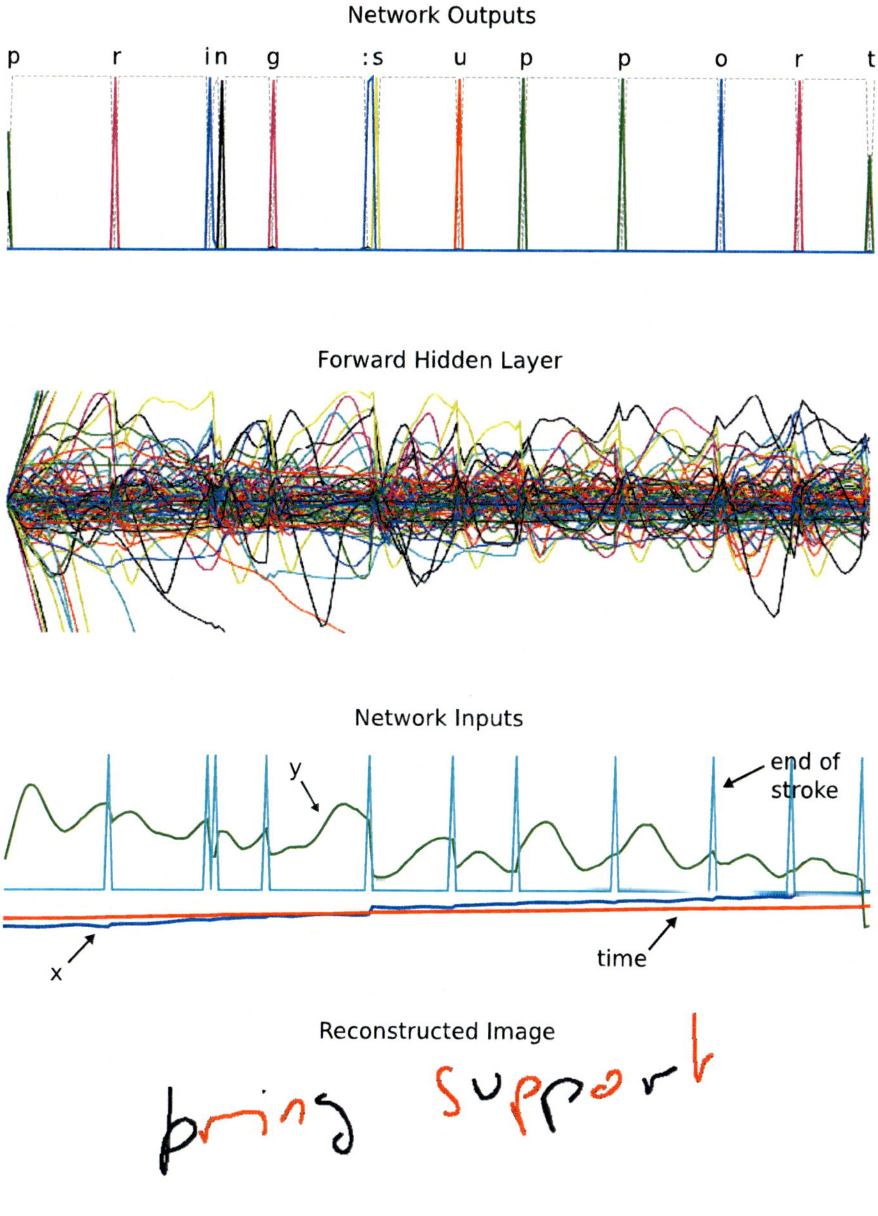


**Fig. 7.9 BLSTM-CTC network labelling an excerpt from IAM-OnDB, using raw inputs.** The ':' label in the outputs is an end-of-word marker. The 'Reconstructed Image' was recreated from the pen positions stored by the sensor. Successive strokes have been alternately coloured red and black to highlight their boundaries. Note that strokes do not necessarily correspond to individual letters: this is no problem for CTC because it does not require segmented data. This example demonstrates the robustness of CTC to line slope, and illustrates the need for context when classifying letters (the 'ri' in 'bring' is ambiguous on its own, and the final 't' could equally be an 'l').

**Table 7.6 Word recognition on IAM-OnDB.** The error measure is the word error rate. LM = language model. BLSTM-CTC results are a mean over 4 runs, ± standard error. All differences were significant ($p < 0.01$).

| System | Input | LM | Error (%) |
|---|---|---|---|
| HMM | Preprocessed | ✓ | 35.5 |
| BLSTM-CTC | Raw | ✗ | 30.1 ± 0.5 |
| BLSTM-CTC | Preprocessed | ✗ | 26.0 ± 0.3 |
| BLSTM-CTC | Raw | ✓ | 22.8 ± 0.2 |
| BLSTM-CTC | Preprocessed | ✓ | 20.4 ± 0.3 |

reflects the fact that the preprocessed data contains simpler correlations and shorter time-dependencies, and is therefore easier for the network to learn. It is interesting to note how large the variance in training epochs was for the raw data, given how small the variance in final performance was.

#### 7.6.4.4 Analysis

The results with the raw inputs, where the information required to identify each character is distributed over many timesteps, demonstrate the ability of BLSTM to make use of long range contextual information. An indication of the amount of context required is given by the fact that when we attempted to train a CTC network with a standard BRNN architecture on the same task, it did not converge.

Figures 7.10 and 7.11 show sequential Jacobians for BLSTM-CTC networks using respectively the raw and preprocessed inputs for a phrase from IAM-OnDB. As expected, the size of the region of high sensitivity is considerably larger for the raw representation, because the preprocessing creates localised input features that do not require as much use of long range context.

### *7.6.5 Offline Handwriting Recognition*

This section describes an *offline* handwriting recognition experiment, using the IAM-DB offline handwriting corpus (Marti and Bunke, 2002)[2]. Offline handwriting differs from online in that only the final image created by the pen is available to the algorithm. This makes the extraction of relevant input features more difficult, and usually leads to lower recognition rates.

The images were transformed into one-dimensional feature sequences before being presented to the network. Chapter 9 describes a network capable of directly transcribing offline handwriting images, without requiring any preprocessing.

[2] Available for public download at http://www.iam.unibe.ch/~fki/iamDB

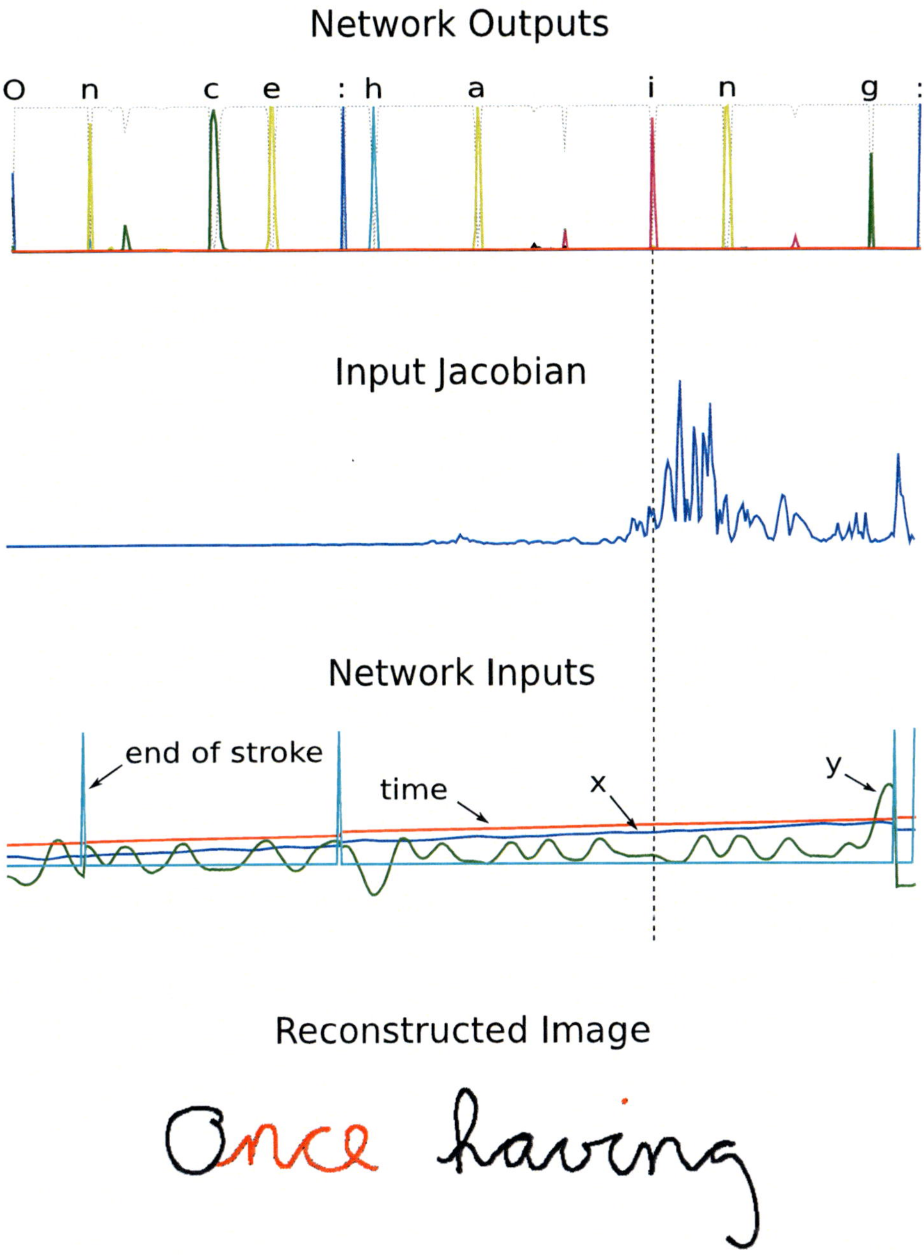


**Fig. 7.10 BLSTM-CTC Sequential Jacobian from IAM-OnDB with raw inputs.** For ease of visualisation, only the derivative with highest absolute value is plotted at each timestep. The Jacobian is plotted for the output corresponding to the label 'i' at the point when 'i' is emitted (indicated by the vertical dashed lines). Note that the network is mostly sensitive to the end of the word: this is possibly because 'ing' is a common suffix, and finding the 'n' and 'g' therefore increases the probability of identifying the 'i'. Note also the spike in sensitivity at the very end of the sequence: this corresponds to the delayed dot of the 'i'.

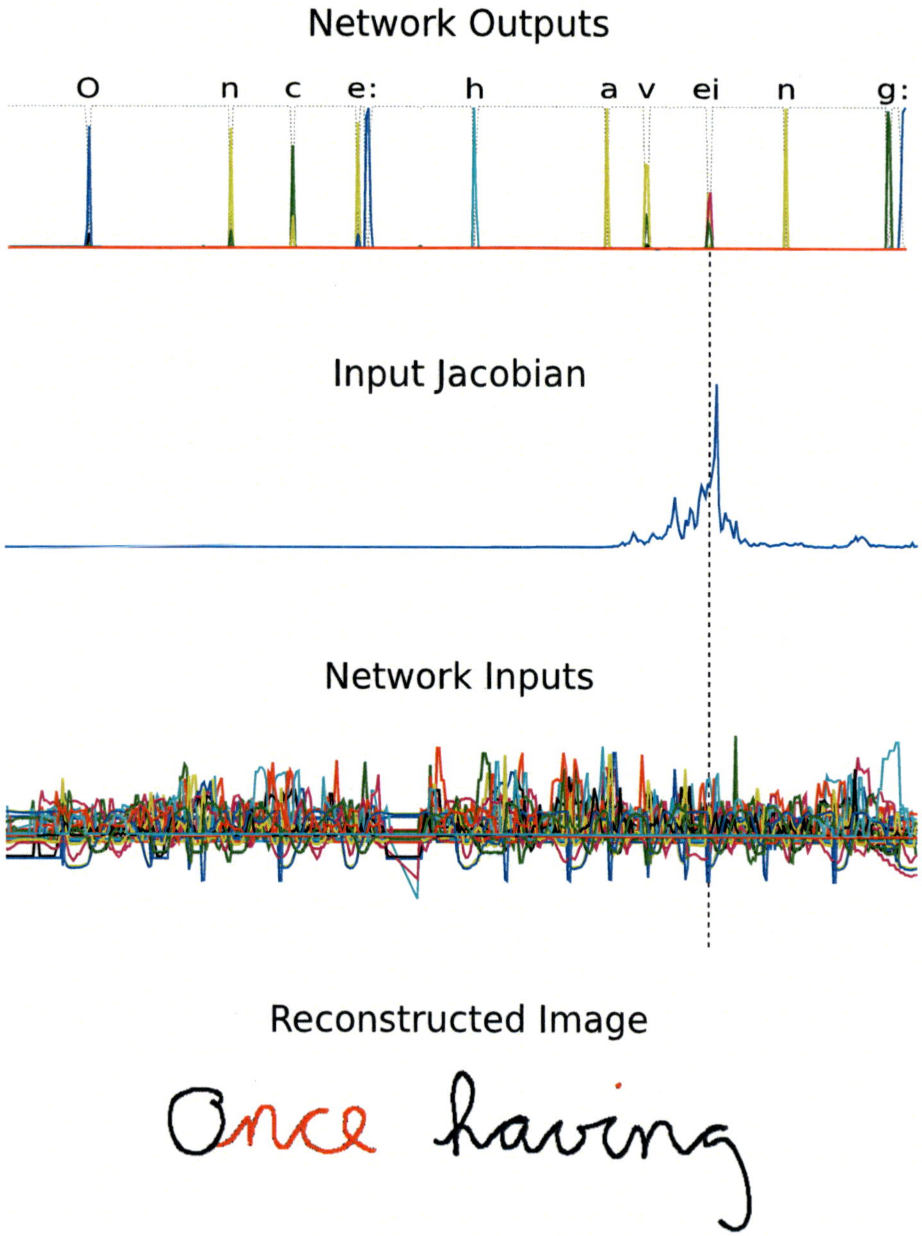


**Fig. 7.11 BLSTM-CTC Sequential Jacobian from IAM-OnDB with pre-processed inputs.** As before, only the highest absolute derivatives are shown, and the Jacobian is plotted at the point when 'i' is emitted. The range of sensitivity is smaller and more symmetrical than for the raw inputs.

#### 7.6.5.1 Data and Preprocessing

The IAM-DB training set contains 6,161 text lines written by 283 writers, the validation set contains 920 text lines by 56 writers, and the test set contains 2,781 text lines by 161 writers. No writer in the test set appears in either the training or validation sets.

Substantial preprocessing was used for this task (Marti and Bunke, 2001). Briefly, to reduce the impact of different writing styles, a handwritten text line image is normalised with respect to skew, slant, and baseline position in the preprocessing phase. After these normalisation steps, a handwritten text line is converted into a sequence of feature vectors. For this purpose a sliding window is used which is moved from left to right, one pixel at each step. Nine geometrical features are extracted at each position of the sliding window.

A dictionary and statistical language model, derived from the same three textual corpora as used in Section 7.6.4, were used for decoding (Bertolami and Bunke, 2007). The integration of the language model was optimised on a validation set. As before, the dictionary consisted of the 20,000 most frequently occurring words in the corpora.

#### 7.6.5.2 Experimental Setup

The HMM-based recogniser was identical to the one used in (Bertolami and Bunke, 2007), with each character modelled by a linear HMM. The number of states was chosen individually for each character (Zimmermann et al., 2006b), and twelve Gaussian mixture components were used to model the output distribution in each state.

The network contained 100 LSTM blocks in the forward and backward hidden layers. There were 9 inputs and 82 outputs, giving 105,082 weights in total. No noise was added during training.

#### 7.6.5.3 Results

From Table 7.7 it can be seen that BLSTM-CTC with a language model gave a mean word error rate of 25.9%, compared to 35.5% with a benchmark HMM. This is an error reduction of 27.0%. Without the language model however, the network did not significantly outperform the HMM. The character error rate for BLSTM-CTC was $18.2 \pm 0.6\%$.

The mean training time for the network was $71.3 \pm 7.5$ epochs.

## 7.7 Discussion

For most of the experiments in this chapter, the performance gap between BLSTM-CTC networks and HMMs is substantial. In what follows, we discuss the key differences between the two systems, and suggest reasons for the network's superiority.

**Table 7.7 Word recognition results on IAM-DB.** The error measure is the word error rate. LM = language model. BLSTM-CTC results are a mean over 4 runs, ± standard error.

| System | LM | Error (%) |
|---|---|---|
| HMM | ✓ | 35.5 |
| BLSTM-CTC | ✗ | 34.6 ± 1.1 |
| BLSTM-CTC | ✓ | 25.9 ± 0.8 |

Firstly, HMMs are generative, while an RNN trained with CTC is discriminative. As discussed in Section 2.2.3, the advantages of the generative approach include the ability to add extra models to an already trained system, and the ability to generate synthetic data. However, discriminative methods tend to give better results for classification tasks, because they focus entirely on finding the correct labels. Additionally, for tasks where the prior data distribution is hard to determine, generative approaches can only provide unnormalised likelihoods for the label sequences. Discriminative approaches, on the other hand, yield normalised label probabilities, which can be used to assess prediction confidence, or to combine the outputs of several classifiers.

A second difference is that RNNs, and particularly LSTM, provide more flexible models of the input features than the mixtures of diagonal Gaussians used in standard HMMs. In general, mixtures of Gaussians can model complex, multi-modal distributions; however, when the Gaussians have diagonal covariance matrices (as is usually the case) they are limited to modelling distributions over independent variables. This assumes that the input features are decorrelated, which can be difficult to ensure for real-world tasks. RNNs, on the other hand, do not assume that the features come from a particular distribution, or that they are independent, and can model nonlinear relationships among features. However, RNNs generally perform better using input features with simpler inter-dependencies.

A third difference is that the internal states of a standard HMM are discrete and single valued, while those of an RNN are defined by the vector of activations of the hidden units, and are therefore continuous and multivariate. This means that for an HMM with $N$ states, only $\mathcal{O}(logN)$ bits of information about the past observation sequence are carried by the internal state. For an RNN, on the other hand, the amount of internal information grows linearly with the number of hidden units.

A fourth difference is that HMMs are constrained to segment the input sequence in order to determine the sequence of hidden states. This is often problematic for continuous input sequences, since the precise boundary between units, such as cursive characters, can be ambiguous. It is also an unnecessary burden in tasks such as keyword spotting, where most of the inputs should be ignored. A further problem with segmentation is that, at least with standard Markovian transition probabilities, the probability of remaining in a particular state decreases exponentially with time. Exponential decay is in general a poor model of state duration, and various measures have been suggested to alleviate this (Johnson, 2005). However, an RNN trained with CTC does not need to segment the input sequence, and therefore avoids both of these problems.

A final, and perhaps most crucial, difference is that unlike RNNs, HMMs assume the probability of each observation to depend only on the current state. A consequence of this is that data consisting of continuous trajectories (such as the sequence of pen coordinates for online handwriting, and the sequence of window positions in offline handwriting) are difficult to model with standard HMMs, since each observation is heavily dependent on those around it. Similarly, data with long range contextual dependencies is troublesome, because individual sequence elements (such as letters or phonemes) are influenced by the elements surrounding them. The latter problem can be mitigated by adding extra models to account for each sequence element in all different contexts (e.g., using triphones instead of phonemes for speech recognition). However, increasing the number of models exponentially increases the number of parameters that must be inferred which, in turn, increases the amount of data required to reliably train the system. For RNNs on the other hand, modelling continuous trajectories is natural, since their own hidden state is itself a continuous trajectory. Furthermore, the range of contextual information accessible to an RNN during a particular output prediction can in principle extend to the entire input sequence.

# Chapter 8
# Multidimensional Networks

Recurrent neural networks are an effective architecture for sequence learning tasks where the data is strongly correlated along a single axis. This axis typically corresponds to time, or in some cases (such as protein secondary structure prediction) one-dimensional space. Some of the properties that make RNNs suitable for sequence learning, such as robustness to input warping and the ability to learn which context to use, are also desirable in domains with more than one spatio-temporal dimension. However, standard RNNs are inherently one dimensional, and therefore poorly suited to multidimensional data. This chapter describes multidimensional recurrent neural networks (MDRNNs; Graves et al., 2007), a special case of directed acyclic graph RNNs (DAG-RNNs; Baldi and Pollastri, 2003), extend the potential applicability of RNNs to vision, video processing, medical imaging and many other areas, while avoiding the scaling problems that have plagued other multidimensional models. We also introduce multidimensional Long Short-Term Memory, thereby bringing the benefits of long range contextual processing to multidimensional tasks.

Although we will focus on the application of MDRNNs to supervised labelling and classification, it should be noted that the same architecture could be used for a wide range of tasks requiring the processing of multidimensional data.

Section 8.1 provides the background material and literature review for multidimensional algorithms. Section 8.2 describes the MDRNN architecture in detail. Section 8.3 presents experimental results on two image classification tasks.

## 8.1 Background

Recurrent neural networks were originally developed as a way of extending feedforward neural networks to sequential data. The addition of recurrent connections allows RNNs to make use of previous context, as well as making them more more robust to warping along the time axis than non-recursive models. Access to contextual information and robustness to warping are also

A. Graves: Supervised Sequence Labell. with Recur. Neur. Networks, SCI 385, pp. 95–108.
springerlink.com 

important when dealing with multidimensional data. For example, a face recognition algorithm should use the entire face as context, and should be robust to changes in perspective, distance etc. It therefore seems desirable to apply RNNs to such tasks.

However, the standard RNN architectures are inherently one dimensional, meaning that in order to use them for multidimensional tasks, the data must be preprocessed to one dimension, for example by presenting one vertical line of an image at a time to the network. Perhaps the best known use of neural networks for multidimensional data has been the application of convolutional networks (LeCun et al., 1998a) to image processing tasks such as digit recognition (Simard et al., 2003). One disadvantage of convolutional networks is that, because they are not recurrent, they rely on hand specified kernel sizes to introduce context. Another disadvantage is that they do not scale well to large images. For example, sequences of handwritten digits must be presegmented into individual characters before they can be recognised by convolutional networks (LeCun et al., 1998a).

Other neural network based approaches to two-dimensional data (Pang and Werbos, 1996; Lindblad and Kinser, 2005) have been structured like cellular automata. A network update is performed at every timestep for every data-point, and contextual information propagates one step at a time in all directions. This provides an intuitive solution to the problem of simultaneously assimilating context from all directions. However the computational cost of accessing long-range context is a serious drawback: if the data contains a total of $T$ points then $\mathcal{O}(NT)$ weight updates are required to access information $N$ steps away.

A more efficient way of building multidimensional context into recurrent networks is provided by *directed acyclic graph RNNs* (DAG-RNNs; Baldi and Pollastri, 2003; Pollastri et al., 2006). DAG-RNNs are networks whose update order is determined by an arbitrary directed acyclic graph. The fact that the update graph is acyclic ensures that the network can process the entire sequence in one pass, making the diffusion of context information as efficient as it is for ordinary RNNs. Various forms of DAG-RNN appeared in publications prior to Baldi's (Goller, 1997; Sperduti and Starita, 1997; Frasconi et al., 1998; Hammer, 2002). However these works mostly focused on networks with tree-structured update graphs, whereas we are interested in the case where the graph corresponds to an n-dimensional grid. If $2^n$ distinct hidden layers are used to process the grid along all possible directions, and all of these layers feed forward to the same output layer, the resulting network is termed the 'canonical' n-dimensional generalisation of bidirectional RNNs by Baldi. In two dimensions, this structure has been successfully used to evaluate positions in the board game 'Go' (Wu and Baldi, 2006) and to determine two dimensional protein contact maps (Baldi and Pollastri, 2003). The multidimensional networks discussed in this chapter are equivalent to n-dimensional canonical DAG-RNNs, although the formulation is somewhat different.

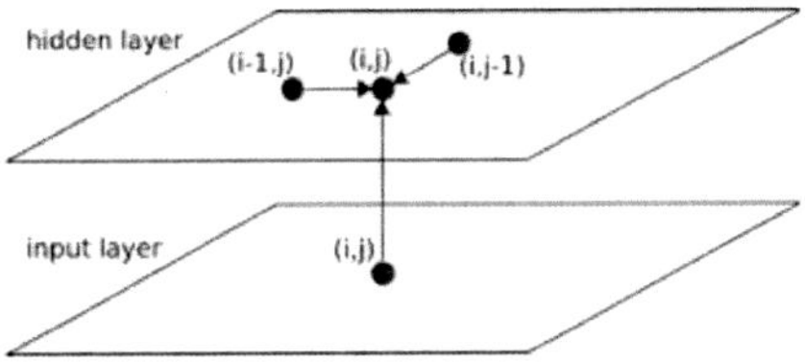


**Fig. 8.1 MDRNN forward pass**

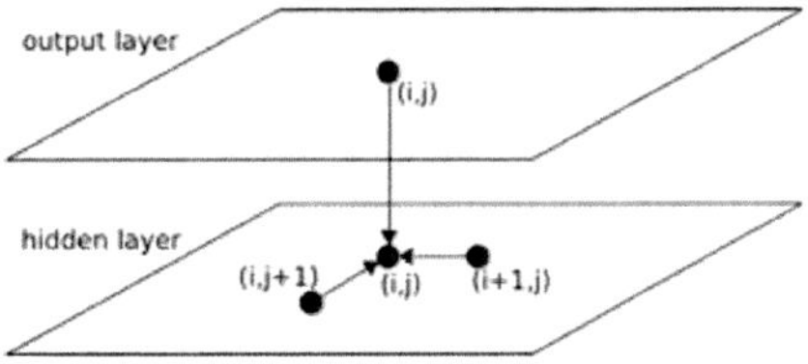


**Fig. 8.2 MDRNN backward pass**

Various statistical models have been proposed for multidimensional data, notably multidimensional hidden Markov models. However, multidimensional HMMs suffer from two serious drawbacks: (1) the time required to run the Viterbi algorithm, and thereby calculate the optimal state sequences, grows exponentially with the size of the data exemplars, and (2) the number of transition probabilities, and hence the required memory, grows exponentially with the data dimensionality. Numerous approximate methods have been proposed to alleviate one or both of these problems, including pseudo 2D and 3D HMMs (Hülsken et al., 2001), isolating elements (Li et al., 2000), approximate Viterbi algorithms (Joshi et al., 2005), and dependency tree HMMs (Jiten et al., 2006). However, none of these methods exploit the full multidimensional structure of the data.

As we will see, MDRNNs bring the benefits of RNNs to multidimensional data, without suffering from the scaling problems described above.

## 8.2 Network Architecture

The basic idea of MDRNNs is to replace the single recurrent connection found in standard RNNs with as many recurrent connections as there are dimensions in the data. During the forward pass, at each point in the data sequence, the hidden layer of the network receives both an external input and its own activations from one step back along all dimensions. Figure 8.1 illustrates the two dimensional case.

Note that, although the word *sequence* usually denotes one dimensional data, we will use it to refer to independent data exemplars of any dimensionality. For example, an image is a two dimensional sequence, a video is a three dimensional sequence, and a series of fMRI brain scans is a four dimensional sequence.

Clearly, the data must be processed in such a way that when the network reaches a point in an n-dimensional sequence, it has already passed through all the points from which it will receive its previous activations. This can be ensured by following a suitable ordering on the set of points $\{(p_1, p_2, ..., p_n)\}$. One example of such an ordering is $(p_1, \ldots, p_n) < (p'_1, \ldots, p'_n)$ if $\exists\ m \in (1, \ldots, n)$ such that $p_m < p'_m$ and $p_d = p'_d\ \forall\ d \in (1, \ldots, m-1)$. Note that this is not the only possible ordering, and that its realisation for a particular sequence depends on an arbitrary choice of axes. We will return to this point

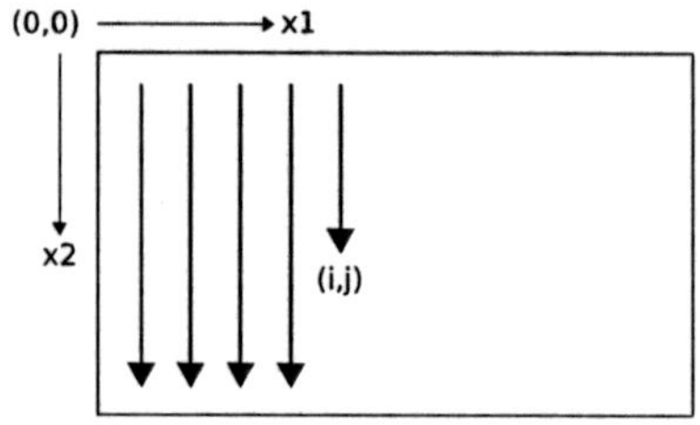


**Fig. 8.3 Sequence ordering of 2D data.** The MDRNN forward pass starts at the origin and follows the direction of the arrows. The point (i,j) is never reached before both (i-1,j) and (i,j-1).

in Section 8.2.1. Figure 8.3 illustrates the above ordering for a 2 dimensional sequence.

The forward pass of an MDRNN can then be carried out by feeding forward the input and the $n$ previous hidden layer activations at each point in the ordered input sequence, and storing the resulting hidden layer activations. Care must be taken at the sequence boundaries not to feed forward activations from points outside the sequence.

Note that the 'points' in the input sequence will in general be multivalued vectors. For example, in a two dimensional colour image, the inputs could be single pixels represented by RGB triples, or groups of pixels, or features provided by a preprocessing method such as a discrete cosine transform.

The error gradient of an MDRNN (that is, the derivative of some loss function with respect to the network weights) can be calculated with an n-dimensional extension of backpropagation through time (BPTT; see Section 3.1.4 for more details). As with one dimensional BPTT, the sequence is processed in the reverse order of the forward pass. At each timestep, the hidden layer receives both the output error derivatives and its own $n$ 'future' derivatives. Figure 8.2 illustrates the BPTT backward pass for two dimensions. Again, care must be taken at the sequence boundaries.

Define $a_j^{\mathbf{p}}$ and $b_j^{\mathbf{p}}$ respectively as the network input to unit $j$ and the activation of unit $j$ at point $\mathbf{p} = (p_1, \ldots, p_n)$ in an n-dimensional sequence $\mathbf{x}$. Let $w_{ij}^d$ be the weight of the recurrent connection from unit $i$ to unit $j$ along dimension $d$. Consider an n-dimensional MDRNN with $I$ input units, $K$ output units, and $H$ hidden summation units. Let $\theta_h$ be the activation function of hidden unit $h$. Then the forward pass up to the hidden layer for a sequence with dimensions $(D_1, D_2, \ldots, D_n)$ is given in Algorithm 8.1.

Note that units are indexed starting at 1, while coordinates are indexed starting at 0. For some unit $j$ and some differentiable loss function $\mathcal{L}$, define

$$\delta_j^{\mathbf{p}} \stackrel{\text{def}}{=} \frac{\partial \mathcal{L}}{\partial a_j^{\mathbf{p}}} \tag{8.1}$$

Then the backward pass for the hidden layer is given in Algorithm 8.2.

**for** $p_1 = 0$ to $D_1 - 1$ **do**
  **for** $p_2 = 0$ to $D_2 - 1$ **do**
    ...
    **for** $p_n = 0$ to $D_n - 1$ **do**
      **for** $h = 1$ to $H$ **do**
        $a_h^{\mathbf{p}} = \sum_{i=1}^{I} x_i^{\mathbf{p}} w_{ih}$
        **for** $d = 1$ to $n$ **do**
          **if** $p_d > 0$ **then**
            $a_h^{\mathbf{p}} \mathrel{+}= \sum_{h'=1}^{H} b_{h'}^{(p_1,\ldots,p_d-1,\ldots,p_n)} w_{h'h}^d$
        $b_h^{\mathbf{p}} = \theta_h(a_h^{\mathbf{p}})$

**Algorithm 8.1** MDRNN Forward Pass

**for** $p_1 = D_1 - 1$ to 0 **do**
  **for** $p_2 = D_2 - 1$ to 0 **do**
    ...
    **for** $p_n = D_n - 1$ to 0 **do**
      **for** $h = 1$ to $H$ **do**
        $e_h^{\mathbf{p}} = \sum_{k=1}^{K} \delta_k^{\mathbf{p}} w_{hk}$
        **for** $d = 1$ to $n$ **do**
          **if** $p_d < D_d - 1$ **then**
            $e_h^{\mathbf{p}} \mathrel{+}= \sum_{h'=1}^{H} \delta_{h'}^{(p_1,\ldots,p_d+1,\ldots,p_n)} w_{hh'}^d$
        $\delta_h^{\mathbf{p}} = \theta'_h(e_h^{\mathbf{p}})$

**Algorithm 8.2** MDRNN Backward Pass

Defining $\mathbf{p}_d^- \stackrel{\text{def}}{=} (p_1, \ldots, p_d - 1, \ldots, p_n)$ and $\mathbf{p}_d^+ \stackrel{\text{def}}{=} (p_1, \ldots, p_d + 1, \ldots, p_n)$, the above procedures can be compactly expressed as follows:

### Forward Pass

$$a_h^{\mathbf{p}} = \sum_{i=1}^{I} x_i^{\mathbf{p}} w_{ih} + \sum_{\substack{d=1:\\ p_d>0}}^{n} \sum_{h'=1}^{H} b_{h'}^{\mathbf{p}_d^-} w_{h'h}^d \tag{8.2}$$

$$b_h^{\mathbf{p}} = \theta_h(a_h^{\mathbf{p}}) \tag{8.3}$$

### Backward Pass

$$\delta_h^{\mathbf{p}} = \theta'_h(a_h^{\mathbf{p}}) \left( \sum_{k=1}^{K} \delta_k^{\mathbf{p}} w_{hk} + \sum_{\substack{d=1:\\ p_d<D_d-1}}^{n} \sum_{h'=1}^{H} \delta_{h'}^{\mathbf{p}_d^+} w_{hh'}^d \right) \tag{8.4}$$

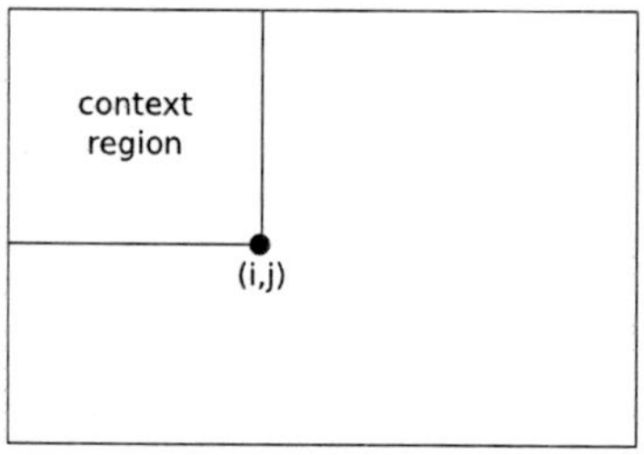


**Fig. 8.4 Context available at (i,j) to a unidirectional MDRNN**

Since the forward and backward pass require one pass each through the data sequence, the overall complexity of MDRNN training is linear in the number of data points and the number of network weights.

In the special case where $n = 1$, the above equations reduce to those of a standard RNN (Section 3.2).

### *8.2.1 Multidirectional Networks*

At some point $(p_1, \ldots, p_n)$ in the input sequence, the network described above has access to all points $(p'_1, \ldots, p'_n)$ such that $p'_d \leq p_d \; \forall \; d \in (1, \ldots, n)$. This defines an n-dimensional 'context region' of the full sequence, as illustrated in Figure 8.4. For some tasks, such as object recognition, this would in principle be sufficient. The network could process the image according to the ordering, and output the object label at a point when the object to be recognised is entirely contained in the context region.

However it is usually preferable for the network to have access to the surrounding context in all directions. This is particularly true for tasks where precise localisation is required, such as image segmentation. As discussed in Chapter 3, for one dimensional RNNs, the problem of multidirectional context was solved by the introduction of bidirectional recurrent neural networks (BRNNs). BRNNs contain two separate hidden layers that process the input sequence in the forward and reverse directions. The two hidden layers are connected to a single output layer, thereby providing the network with access to both past and future context.

BRNNs can be extended to n-dimensional data by using $2^n$ separate hidden layers, each of which processes the sequence using the ordering defined above, but with a different choice of axes. The axes are chosen so that each one has its origin on a distinct vertex of the sequence. The 2 dimensional case is illustrated in Figure 8.5. As before, the hidden layers are connected to a single output layer, which now has access to all surrounding context (see Figure 8.6). Baldi refers to this structure as the "canonical" generalisation of BRNNs (Baldi and Pollastri, 2003).

Clearly, if the size of the hidden layers is held constant, the complexity of the multidirectional MDRNN architecture scales as $\mathcal{O}(2^n)$ for n-dimensional

#### 8.2.2.1 Forward Pass

*Input Gates*

$$a_\iota^{\mathbf{p}} = \sum_{i=1}^{I} x_i^{\mathbf{p}} w_{i\iota} + \sum_{\substack{d=1:\\ p_d>0}}^{n} \left( \sum_{h=1}^{H} b_h^{\mathbf{p}_d^-} w_{h\iota}^d + \sum_{c=1}^{C} w_{c\iota} s_c^{\mathbf{p}_d^-} \right) \tag{8.5}$$

$$b_\iota^{\mathbf{p}} = f(a_\iota^{\mathbf{p}}) \tag{8.6}$$

*Forget Gates*

$$a_{\phi,d}^{\mathbf{p}} = \sum_{i=1}^{I} x_i^{\mathbf{p}} w_{i(\phi,d)} + \sum_{\substack{d'=1:\\ p_{d'}>0}}^{n} \sum_{h=1}^{H} b_h^{\mathbf{p}_{d'}^-} w_{h(\phi,d)}^{d'} + \begin{cases} \sum_{c=1}^{C} w_{c(\phi,d)} s_c^{\mathbf{p}_d^-} & \text{if } p_d > 0 \\ 0 \text{ otherwise} \end{cases} \tag{8.7}$$

$$b_{\phi,d}^{\mathbf{p}} = f(a_{\phi,d}^{\mathbf{p}}) \tag{8.8}$$

*Cells*

$$a_c^{\mathbf{p}} = \sum_{i=1}^{I} x_i^{\mathbf{p}} w_{ic} + \sum_{\substack{d=1:\\ p_d>0}}^{n} \sum_{h=1}^{H} b_h^{\mathbf{p}_d^-} w_{hc}^d \tag{8.9}$$

$$s_c^{\mathbf{p}} = b_\iota^{\mathbf{p}} g(a_c^{\mathbf{p}}) + \sum_{\substack{d=1:\\ p_d>0}}^{n} s_c^{\mathbf{p}_d^-} b_{\phi,d}^{\mathbf{p}} \tag{8.10}$$

*Output Gates*

$$a_\omega^{\mathbf{p}} = \sum_{i=1}^{I} x_i^{\mathbf{p}} w_{i\omega} + \sum_{\substack{d=1:\\ p_d>0}}^{n} \sum_{h=1}^{H} b_h^{\mathbf{p}_d^-} w_{h\omega}^d + \sum_{c=1}^{C} w_{c\omega} s_c^{\mathbf{p}} \tag{8.11}$$

$$b_\omega^{\mathbf{p}} = f(a_\omega^{\mathbf{p}}) \tag{8.12}$$

*Cell Outputs*

$$b_c^{\mathbf{p}} = b_\omega^{\mathbf{p}} h(s_c^{\mathbf{p}}) \tag{8.13}$$

#### 8.2.2.2 Backward Pass

$$\epsilon_c^{\mathbf{p}} \stackrel{\text{def}}{=} \frac{\partial \mathcal{L}}{\partial b_c^{\mathbf{p}}} \qquad \epsilon_s^{\mathbf{p}} \stackrel{\text{def}}{=} \frac{\partial \mathcal{L}}{\partial s_c^{\mathbf{p}}} \tag{8.14}$$

*Cell Outputs*

$$\epsilon_c^{\mathbf{p}} = \sum_{k=1}^{K} \delta_k^{\mathbf{p}} w_{ck} + \sum_{\substack{d=1:\\ p_d<D_d-1}}^{n} \sum_{h=1}^{H} \delta_h^{\mathbf{p}_d^+} w_{ch}^d \tag{8.15}$$

*Output Gates*

$$\delta_\omega^{\mathbf{p}} = f'(a_\omega^{\mathbf{p}}) \sum_{c=1}^{C} \epsilon_c^{\mathbf{p}} h(s_c^{\mathbf{p}}) \tag{8.16}$$

*States*

$$\epsilon_s^{\mathbf{p}} = b_\omega^{\mathbf{p}} h'(s_c^{\mathbf{p}}) \epsilon_c^{\mathbf{p}} + \delta_\omega^{\mathbf{p}} w_{c\omega} + \sum_{\substack{d=1: \\ p_d < D_d - 1}}^{n} \left( \epsilon_s^{\mathbf{p}_d^+} b_{\phi,d}^{\mathbf{p}_d^+} + \delta_\iota^{\mathbf{p}_d^+} w_{c\iota} + \delta_{\phi,d}^{\mathbf{p}_d^+} w_{c(\phi,d)} \right) \tag{8.17}$$

*Cells*

$$\delta_c^{\mathbf{p}} = b_\iota^{\mathbf{p}} g'(a_c^{\mathbf{p}}) \epsilon_s^{\mathbf{p}} \tag{8.18}$$

*Forget Gates*

$$\delta_{\phi,d}^{\mathbf{p}} = \begin{cases} f'(a_{\phi,d}^{\mathbf{p}}) \sum_{c=1}^{C} s_c^{\mathbf{p}_d^-} \epsilon_s^{\mathbf{p}} \text{ if } p_d > 0 \\ 0 \text{ otherwise} \end{cases} \tag{8.19}$$

*Input Gates*

$$\delta_\iota^{\mathbf{p}} = f'(a_\iota^{\mathbf{p}}) \sum_{c=1}^{C} g(a_c^{\mathbf{p}}) \epsilon_s^{\mathbf{p}} \tag{8.20}$$

## 8.3 Experiments

### *8.3.1 Air Freight Data*

The Air Freight database (McCarter and Storkey, 2007) is a ray-traced colour image sequence that comes with a ground truth segmentation into the different textures mapped onto the 3-d models (Figure 8.7). The sequence is 455 frames long and contains 155 distinct textures. Each frame is 120 pixels high and 160 pixels wide.

The advantage of ray-traced data is that the true segmentation can be determined directly from the 3D models. Although the images are not real, they are at least somewhat realistic with shadows, reflections and highlights that make the segmentation challenging to determine.

We used the individual frames in the video sequence to define a 2D image segmentation task, where the aim was to assign each pixel in the input data to the correct texture class. We divided the data at random into a 250 frame train set, a 150 frame test set and a 55 frame validation set. We could instead have defined a 3D task where the network processed segments of the video as independent sequences. However, this would have left us with fewer examples for training and testing.

**Fig. 8.7 Frame from the Air Freight database.** The original image is on the left and the colour-coded texture segmentation is on the right.

For this task we used a multidirectional 2D RNN with LSTM hidden layers. Each of the 4 layers consisted of 25 memory blocks, each containing 1 cell, 2 forget gates, 1 input gate, 1 output gate and 5 peephole weights. This gave a total 600 hidden units. The input and output activation function of the cells was tanh, and the activation function for the gates was the logistic sigmoid. The input layer was size 3 (one each for the red, green and blue components of the pixels) and the output layer was size 155 (one unit for each texture). The network contained 43,257 trainable weights in total. The softmax activation function was used at the output layer, with the cross-entropy loss function (Section 3.1.3). The network was trained using online gradient descent (weight updates after every training sequence) with a learning rate of $10^{-6}$ and a momentum of 0.9.

The final pixel classification error rate, after 330 training epochs, was 7.1% on the test set.

### *8.3.2 MNIST Data*

The MNIST database (LeCun et al., 1998a) of isolated handwritten digits is a subset of a larger database available from NIST. It consists of size-normalised, centred images, each of which is 28 pixels high and 28 pixels wide and contains a single handwritten digit. The data comes divided into a training set with 60,000 images and a test set with 10,000 images. We used 10,000 of the training images for validation, leaving 50,000 for training.

The task on MNIST is to label the images with the corresponding digits. This is a widely-used benchmark for pattern classification algorithms.

We trained the network to perform a slightly modified task where each pixel was classified according to the digit it belonged to, with an additional class for background pixels. We then recovered the original task by choosing for each sequence the digit whose corresponding output unit had the highest cumulative activation over the entire sequence.

To test the network's robustness to input warping, we also evaluated it on an altered version of the MNIST test set, where elastic deformations had been

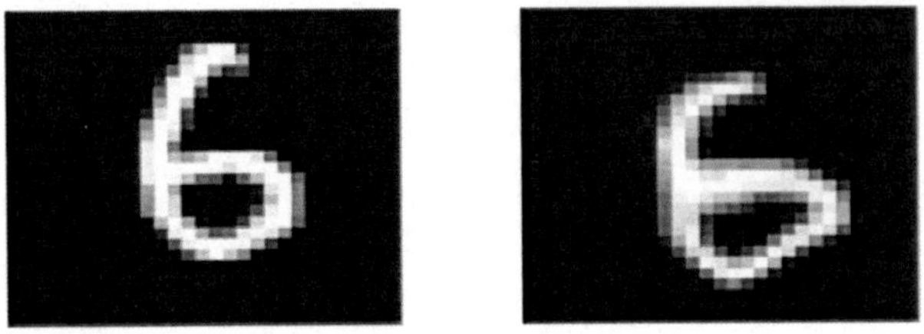

**Fig. 8.8 MNIST image before and after deformation**

**Table 8.1 Classification results on MNIST.** The error measure is the image misclassification rate

| Network | Clean Error (%) | Warped Error (%) |
|---|---|---|
| Convolutional | 0.9 | 11.3 |
| MDRNN | 0.9 | 7.1 |

applied to every image (see Figure 8.8). The deformations were the same as those recommended by Simard (2003) to augment the MNIST training set, with parameters $\sigma = 4.0$ and $\alpha = 34.0$, and using a different initial random field for every sample image.

We compared our results with the convolutional neural network that has achieved the best results so far on MNIST (Simard et al., 2003). Note that we re-implemented the convolutional network ourselves, and we did not augment the training set with elastic distortions, which gives a substantial improvement in performance.

The MDRNN for this task was identical to that for the Air Freight task with the following exceptions: the sizes of the input and output layers were now 1 (for grayscale pixels) and 11 (one for each digit, plus background) respectively, giving 27,511 weights in total, and the learning rate was $10^{-5}$.

Table 8.1 shows that the MDRNN matched the convolutional network on the clean test set, and was considerably better on the warped test set. This suggests that MDRNNs are more robust to input warping than convolutional networks. For the MDRNN, the *pixel* classification error rates (as opposed to the *image* classification error rates) were 0.4% on the clean test set and 3.8% on the warped test set.

One area in which the convolutional net greatly outperformed the MDRNN was training time. The MDRNN required 95 training epochs to converge, whereas the convolutional network required 20. Furthermore, each training epoch took approximately 3.7 hours for the MDRNN compared to under ten minutes for the convolutional network, using a similar processor. The total training time for the MDRNN was over two weeks.

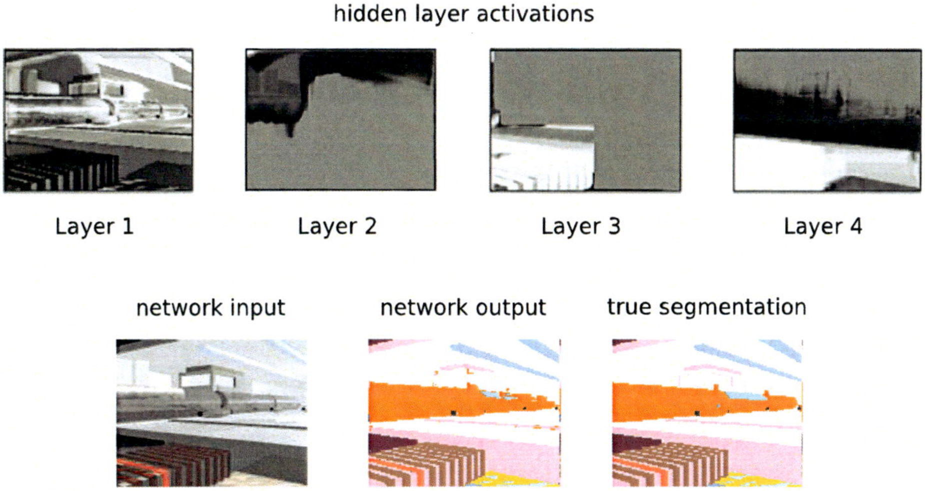


**Fig. 8.9 MDRNN applied to an image from the Air Freight database.** The hidden layer activations display one unit from each of the layers. A common behaviour is to 'mask off' parts of the image, exhibited here by layers 2 and 3.

### *8.3.3 Analysis*

One benefit of two dimensional tasks is that the operation of the network can be easily visualised. Figure 8.9 shows the network activations during a frames from the Air Freight database. As can be seen, the network segments this image almost perfectly, in spite of difficult, reflective surfaces such as the glass and metal tube running from left to right. Clearly, classifying individual pixels in such a surface requires the use of contextual information.

Figure 8.10 shows the absolute value of the sequential Jacobian of an output during classification of an image from the MNIST database. It can be seen that the network responds to context from across the entire image, and

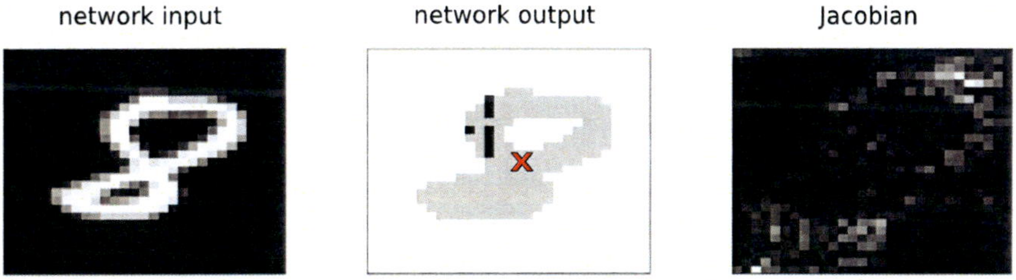


**Fig. 8.10 Sequential Jacobian of an MDRNN for an image from MNIST.** The white outputs correspond to the class 'background' and the light grey ones to '8'. The black outputs represent misclassifications. The output pixel for which the Jacobian is calculated is marked with a cross. Absolute values are plotted for the Jacobian, and lighter colours are used for higher values.

seems particularly attuned to the outline of the digit. Note that the high sensitivity to the very corners of the image is irrelevant, since these pixels are always black in MNIST; this illustrates the need for caution when interpreting the sequential Jacobian.

# Chapter 9
# Hierarchical Subsampling Networks

So far we have focused on recurrent neural networks with a single hidden layer (or set of disconnected hidden layers, in the case of bidirectional or multidirectional networks). As discussed in Section 3.2, this structure is in principle able to approximate any sequence-to-sequence function arbitrarily well, and should therefore be sufficient for any sequence labelling task. In practice however, it tends to struggle with very long sequences. One problem is that, because the entire network is activated at every step of the sequence, the computational cost can be prohibitively high. Another is that the information tends to be more spread out in longer sequences, and sequences with longer range interdependencies are generally harder to learn from.

The effect of sequence length is particularly apparent when the same data is represented in different ways. For example, in the speech recognition experiments we have considered so far the audio data has been pre-processed into sequences consisting of one feature vector for every 10ms of audio. Were we to use raw audio data instead, with a sampling rate of, say, 48KHz, each utterance would be 480 times as long, and the network would therefore need to have approximately 1/480 times the number of weights to process the data at the same speed. Clearly we could not expect comparable performance from such a network. And even given an equally large network, the typical spacing between related inputs would be 480 times as long, which would greatly increase the demands on the network's memory.

A common way to reduce the length of data sequences is to *subsample* them: that is, gather together consecutive timesteps into blocks or *windows*. Given a hierarchy of sequence processors, where the output of the processor at one level is the input of the processor at the next, we can progressively reduce the length by subsampling each output sequence before passing it up the hierarchy. So-called *hierarchical subsampling* is commonly used in fields such as computer vision where the volume of data is too great to be processed by a 'flat' architecture (LeCun et al., 1998b; Reisenhuber and Poggio, 1999). As well as reducing computational cost, it also reduces the effective dispersal of the data, since inputs that are widely separated at the bottom of the hierarchy are transformed to features that are close together at the top.

A. Graves: Supervised Sequence Labell. with Recur. Neur. Networks, SCI 385, pp. 109–131.
springerlink.com 

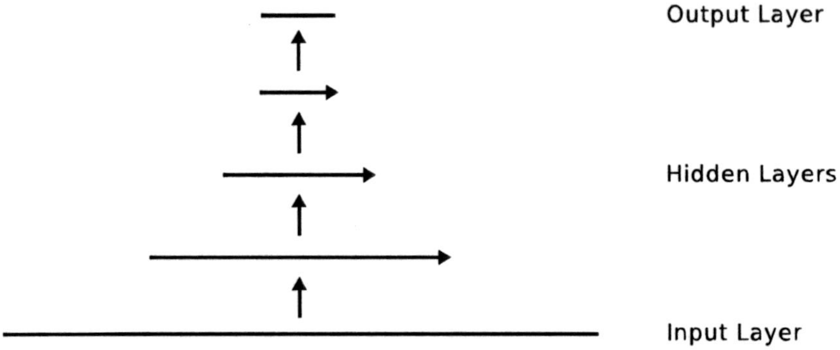


**Fig. 9.1 Information flow through an HSRNN.** The input sequence is subsampled and then scanned by a recurrent hidden layer. The sequence of hidden layer activations is subsampled again and scanned by the next hidden layer. The activations of the final hidden layer are fed without subsampling to the output layer. Note the progressive shortening of the sequence as it moves up the hierarchy.

This chapter introduces *hierarchical subsampling recurrent neural networks* (HSRNNs; Graves and Schmidhuber, 2009) for large data sequences. Although we will focus on the application of HSRNNs to sequence labelling, the architecture is quite general and should be applicable to many sequence learning problems.

Section 9.1 describes the architecture in detail, and Section 9.2 provides experimental results for speech and handwriting recognition.

## 9.1 Network Architecture

A *hierarchical subsampling recurrent neural network* (HSRNN) consists of an input layer, an output layer and multiple levels of recurrently connected hidden layers. The output sequence of each level in the hierarchy is used as the input sequence for the next level up. All input sequences are subsampled using *subsampling windows* of predetermined width, apart from the input to the output layer. The overall flow of information through an HSRNN is sketched in Figure 9.1, while Figure 9.2 provides a more detailed view of an unfolded HSRNN. The structure is similar to that used by convolutional networks (LeCun et al., 1998b), except with recurrent, rather than feedforward, hidden layers.

For each layer in the hierarchy, the forward pass equations are identical to those for a standard RNN (see Section 3.2), except that the sum over input units is replaced by a sum of sums over the subsampling window. For a hidden layer with $H$ units, an 'input' layer (i.e. the next layer down in the hierarchy) with $I$ units, and a size $S$ subsampling window, the hidden activations $b_h^t$ for $1 \leq t \leq T$ can be calculated as follows:

$$a_h^t = \sum_{s=1}^{S}\sum_{i=1}^{I} w_{ih}^s b_i^u + \sum_{h'=1}^{H} w_{h'h} b_{h'}^{t-1} \tag{9.1}$$

$$b_h^t = \theta_h(a_h^t) \tag{9.2}$$

where $w_{ih}^s$ is the weight from input unit $i$ to hidden unit $h$ in step $s$ of the subsampling window, $u = S(t-1)+s$ is the timestep where the window begins in the *unsubsampled* input sequence and $\theta_h$ is a nonlinear activations function (which can be replaced with an LSTM block using the equations in Section 4.6). If the input sequence length is not an exact multiple of the window width, it is padded with zeros at the end. Note that different weights are used for each step of the window, giving a total of $SIH$ weights between the two layers. This provides a more flexible notion of subsampling than is usually applied in sequence processing, where the elements of the subsample window are collapsed to a single vector using a fixed operation (typically an average or maximum).

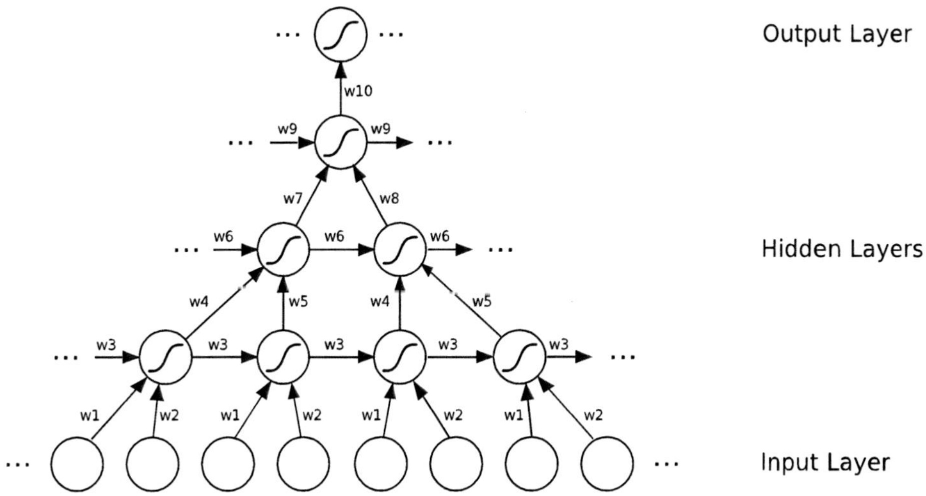


**Fig. 9.2 An unfolded HSRNN.** The same weights are reused for each of the subsampling and recurrent connections along the sequence, giving 10 distinct weight groups (labelled 'w1' to 'w10'). In this case the hierarchy has three hidden level and three subsampling windows, all of size two. The output sequence is one eighth the length of the input sequence.

The complete network is differentiated by first calculating the derivatives of the loss function with respect to the output units (exactly as for normal RNNs) then passing the error gradient backwards through the hierarchy. The gradient for each recurrent layer can be calculated with backpropagation through time, as described in Section 3.2.2, with the $\delta_k^t$ terms provided by the layer above. The gradient must also pass through the subsampling windows between the hidden levels. For a hidden layer with $H$ units, an 'output' layer (i.e. the next layer up in the hierarchy) with $K$ units, the unit derivatives can be calculated as follows:

$$\delta_h^t = \theta_h' \left( \sum_{k=1}^{K} w_{hk}^n \delta_k^s + \sum_{h'=1}^{H} w_{hh'} \delta_{h'}^{t+1} \right) \tag{9.3}$$

where $t$ is in the timestep in the unsubsampled sequence, $n = (t+1) \bmod S$ is the offset within the subsampling window, $s = (t/S)+1$ (with $(t/S)$ rounded down to the nearest integer) is the timestep in the subsampled output sequence and $\delta_i^t \stackrel{\text{def}}{=} \frac{\partial O}{\partial a_i^t}$ as usual. If LSTM is used for the hidden layers, the derivatives $\theta_h'$ with respect to the activation functions should be replaced with the LSTM derivatives given in Section 4.6.

### 9.1.1 Subsampling Window Sizes

Each subsampling operation decreases the length of the output sequence by a factor of the window width. This reduces both the computational cost of the higher levels in the hierarchy, and the effective separation between points in the input sequence. However care must be taken not to make the windows too large. For one thing the output sequence of the network must be long enough for the target sequence, a point we will discuss further in Section 9.1.5. Another issue is that the networks robustness to sequential distortions is incrementally lost as the window size increases. In the limit where the window is as long as the original sequence, a recurrent network reduces to a feedforward network. The windows must therefore be carefully chosen to match the data and the task. For the experiments in this chapter, the basic recipe was to try to keep the window sizes roughly equal at each level (or if necessary, slightly larger at the lower levels), while ensuring that the output sequence was long enough for all the target sequences in the dataset.

Hierarchical subsampling is often carried out with overlapping subsample windows. This has the benefit of including surrounding context in each window. However RNNs are able to supply the context themselves using their recurrent connections, making the overlaps redundant. Furthermore, using overlaps increases both the computational cost and the effective distance between input events—the motivations to use hierarchical subsampling in the first place. Therefore overlapped windows are not used in this book.

### 9.1.2 Hidden Layer Sizes

The hidden layer sizes can be varied independently of the subsampling window sizes. This means that the amount of information in the hidden layer sequences can increase, decrease or remain the same as we move up the hierarchy. More precisely if a hidden layer has $H$ units, the layer below has $I$ units and the subsampling window is size $S$, then the output sequence of the higher layer contains $\frac{HS}{I}$ as many bits as that of the lower layer. This is in contrast to the usual notion of subsampling in signal processing, where the vectors in the subsampled sequence are always the same size as the vectors in the input sequence. However traditional subsampling is primarily carried to to reduce the bit rate of the signal, whereas with HSRNNs we we are more concerned with reducing the lengths of the sequences, and may even want to increase the bit rate.

As with all neural network architectures, the performance of HSRNNs generally increases as the hidden layers get bigger (at least if early stopping or some other regularisation method is used to prevent overfitting). A good rule of thumb is to choose the layer sizes so that each level consumes roughly half the processing time of the level below. For example, if the first hidden layer contains $I$ units (requiring approximately $I^2$ weight operations per timestep), and is subsampled into width $S$ windows, give the next hidden level $H$ units such that

$$H^2 \approx \frac{SI^2}{2} \tag{9.4}$$

This method, which ensures that the total processing time is never much more than twice that of the lowest hidden layer alone, leads to larger layers at the top of the hierarchy than at the bottom. The network therefore progresses from a few high resolution features to many low resolution features, as is typical for hierarchical subsampling systems.

### 9.1.3 Number of Levels

It has been repeatedly noted that training neural networks with many layers using gradient descent is difficult (Hinton et al., 2006; Bengio et al., 2007). This is essentially a restatement of the vanishing gradient problem for RNNs: the sensitivity of the output layer to a given hidden layer tends to decrease the more hidden layers there are between them, just as the sensitivity of an RNN output to a past input decreases as the number of timesteps between them grows. There is therefore a tradeoff between the gain in efficiency and compactness afforded by adding extra hidden levels, and the increased difficulty of training the network. In practice three layers seems to gives good performance for a wide range of data and tasks, and all the experiments in this chapter use three-layer networks.

### 9.1.4 Multidimensional Networks

The extension of HSRNNs to the multidimensional networks covered in the previous chapter is straightforward: the one-dimensional subsampling windows are replaced by multidimensional windows. The sum over timesteps in the subsampling window in (9.1) is therefore replaced with a multidimensional sum over points, and the derivative calculation is modified accordingly. Borrowing the terminology of convolutional networks, we will sometimes refer to the output sequences of two-dimensional network levels as *feature maps*.

#### 9.1.4.1 Multidirectional Networks

Hierarchical subsampling with the *multidirectional* MDRNNs described in Section 8.2.1 (of which bidirectional RNNs—Section 3.2.4—are a special case) is somewhat complicated by the fact that each level of the hierarchy requires $2^n$ hidden layers instead of one. To connect every layer at one level to every layer at the next therefore requires $\mathcal{O}(2^{2n})$ weights. One way to reduce the number of weights is to separate the levels with nonlinear feedforward layers, which reduces the number of weights between the levels to $\mathcal{O}(2^n)$—the same as standard MDRNNs.

The flow of information through a multidirectional HSRNN with feedforward layers is illustrated in Figure 9.3.

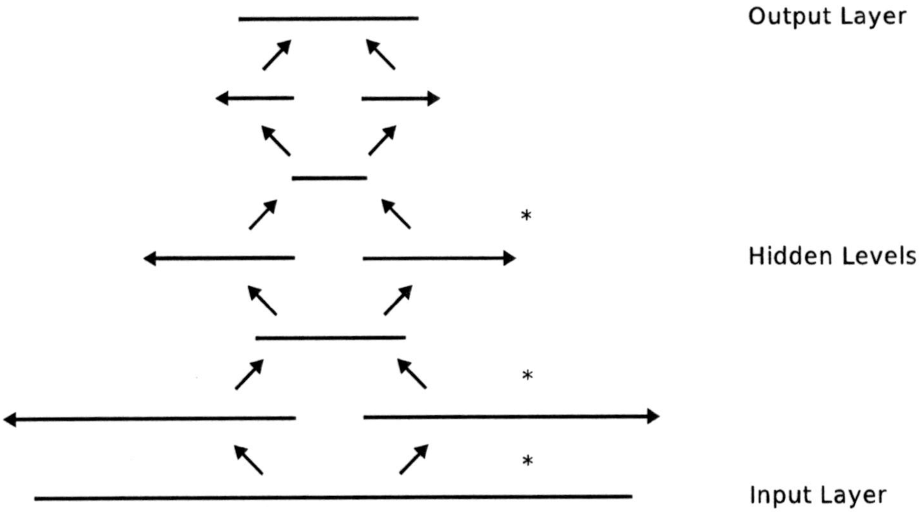


**Fig. 9.3 Information flow through a multidirectional HSRNN.** Each hidden level consists of two recurrent layers scanning in opposite directions. Each pair of hidden levels is separated by a feedforward layer (with no scanning arrow). Subsampling is carries out at the places indicated with a '*'. Note that the outputs of the final hidden level are neither subsampled nor passed through a feedforward layer before being fed to the output layer.

As with the hidden layers, the sizes of the feedforward layers should to be chosen to balance performance against computation time. The feedforward layers act as a bottleneck to the information reaching the higher levels, and making them too small can hamper the network. As a rule of thumb, giving each feedforward layer between half and one times as many units as the combined hidden layers in the level below appears to work well in practice. However we have found it beneficial to tune the precise number of feedforward units separately for every experiment.

Unlike the rest of the network, bias weights are not usually connected to the feedforward layers, because they appear to make no difference to performance (presumably the biases in the recurrent layers are able to compensate).

### 9.1.5 Output Layers

In principle HSRNNs can be trained with the same output layers as ordinary RNNs. However care must be taken to ensure that the output sequence has the correct shape for the corresponding loss function. For example, HSRNNs are clearly unsuited to framewise classification or other tasks where a separate output is required for every input, since this would make subsampling impossible. An HSRNN trained with CTC (Chapter 7), on the other hand, must output a one dimensional sequence at least as long as the target sequence, while an HSRNN used for sequence classification must emit a single output per sequence.

If the dimensions of the input and output sequences are fixed, or even if the relationship between them is fixed (for example if a single classification is required for every fixed-length input segment) the subsample windows can be chosen to ensure that the output sequences are the correct shape. However, the focus of this book is on problems where labels are applied to input patterns of widely varying duration or shape. In what follows, we describe a simple technique for ensuring that the outputs of an HSRNN can be used for sequence and temporal classification.

#### 9.1.5.1 Sequence Classification

Perhaps the most obvious way to classify complete sequences with an RNN is to output a single classification at the end of the sequence. However this is clearly impractical for bidirectional or multidirectional networks, where the sequence 'ends' at different points for different layers. It also requires the network to store the information required to make the classification until the end of the sequence – which may be a long way from where that information was received.

An alternative approach, used in the image classification experiment in Section 8.3.2, is to classify each point in the output sequence independently, then sum up the class probabilities to find the highest ranked class overall. This approach is applicable to multidirectional networks, and allows the network to make localised classifications based on the information it has just received from the input sequence. However it requires the network to make redundant classifications, and is therefore inconsistent with the notion of a neural network as a single-valued function from input sequences to target distributions.

The solution employed in this chapter is to first collapse the output sequence to a single point then classify that. This can be achieved by summing over the inputs to each output unit at all points $\mathbf{p}$ in the output sequence, then applying the softmax function:

$$a_k = \sum_{\mathbf{p}} a_k^{\mathbf{p}} \tag{9.5}$$

$$y_k = \frac{e^{a_k}}{\sum_{\mathbf{p}'} e^{a_{k'}}} \tag{9.6}$$

As well as allowing the network to choose where in the sequence to make predictions, using summation permits different predictions to be made in different places, then weighted against each other in the final classification according to the confidence with which they were made.

#### 9.1.5.2 Connectionist Temporal Classification

The reduction of long output sequences to short target sequences is built into the CTC loss function, and in principle does not have to be provided by the HSRNN. However for extremely long input sequences, such as raw speech data, using subsampling to reduce the output sequence length to within a factor of ten or so of the target sequence length is highly beneficial.

For multidimensional HSRNNs, the output sequence must be collapsed along all but one of its dimensions before CTC can be applied. For example, if the task is to transcribe images of handwritten text, the output sequence should be collapsed vertically (at least for languages with horizontal writing). If the task is to transcribe a video sequence of sign language gestures, the output should be collapsed along the two spatial dimensions, leaving only the time dimension.

If the shape of the input sequence is fixed along the dimensions to be collapsed, a suitable output sequence can be ensured by choosing the right subsample window sizes. In the case of video transcription this is quite feasible, since the spatial dimensions of video data are usually fixed. If the shape of the input sequences is variable, we can use the same summation trick for the output unit activations as we used for sequence classification, only with the sum running over all but one of the output sequence dimensions:

$$a_k^t = \sum_{p_1} \dots \sum_{p_{n-1}} a_k^{\mathbf{p}} \tag{9.7}$$

$$y_k^t = \frac{e^{a_k^t}}{\sum_{\mathbf{p}'} e^{a_{k'}^t}} \tag{9.8}$$

where the $n^{th}$ dimension is the one along which CTC is applied.

### 9.1.6 Complete System

The combination of HSRNNs with a CTC or classification output layer gives a flexible system for labelling large data sequences. The subsampling windows can be adjusted to suit a wide range of input resolutions, and the dimensionality of the network can be chosen to match the dimensionality of the data. Fig. 9.4 illustrates a complete two dimensional, multidirectional HSRNN comprising MDLSTM layers, feedforward layers and a CTC output layer, applied to offline Arabic handwriting recognition.

## 9.2 Experiments

This section assesses the practical efficacy of HSRNNs with experiments on speech and handwriting recognition. the outstanding achievement of HSRNNs so far is winning handwriting recognition competitions in three different languages at the 2009 International Conference on Document Analysis and Recognition. These results are reviewed in Sections 9.2.1 to 9.2.4, while Section 9.2.5 presents phoneme recognition results on the TIMIT database using three different different representations of acoustic data.

HSRNNs contain considerably more hand-tuned parameters (subsampling window sizes, multiple hidden layer sizes etc.) than the RNNs we have considered before. However the majority of these can be held constant for a wide class of sequence labelling tasks. In what follows we distinguish three different types of parameter: those that are fixed for all networks in the chapter; those that are automatically determined by the task, data or by other parameters; and those that are hand-tuned for each network. Only the parameters in the latter two categories will be specified for individual networks.

### *Fixed Parameters*

- The hierarchy contained three levels.
- The three levels were separated by two feedforward layers with the tanh activation function.
- Subsampling windows were applied in three places: to the input sequence, to the output sequence of the first hidden level, and to the output sequence of the second hidden level.
- The network architecture was bidirectional LSTM (Section 4.5) for all experiments with one-dimensional data, and multidirectional MDLSTM (Section 8.2.1) for all experiments with two-dimensional data. Each level of the hierarchy therefore contained 2 hidden layers for one-dimensional data, and 4 hidden layers for two-dimensional data.
- The hidden layers were recurrently connected (all input units connected to all hidden units, all hidden units connected to all output units and all hidden units)
- The LSTM blocks contained one cell each.
- The LSTM gate activation function ($f$ in Figure 4.2) was the logistic sigmoid: $f(x) = 1/(1 + e^{-x})$, while the cell input and output functions ($g$ and $h$ in Figure 4.2) were both tanh.
- Online steepest descent was used for training with a momentum of 0.9 and a learning rate of $1e-4$.

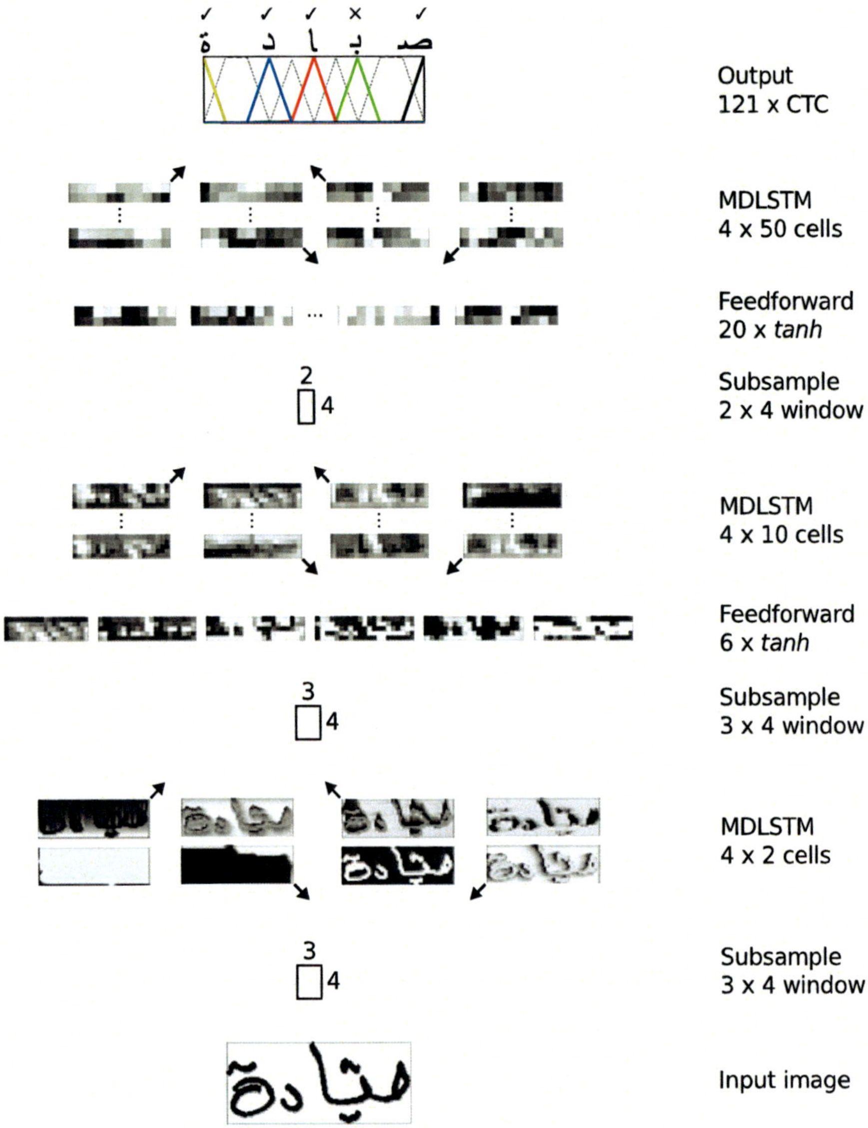


**Fig. 9.4 HSRNN applied to offline Arabic handwriting recognition.** The input image, which consists of a single handwritten Arabic word, is subsampled with a window three pixels wide and four pixels high. The subsampled image is then scanned by four MDLSTM layers, each containing two cells. The feature maps corresponding to the activations of the cells in each layer are displayed, with the arrows in the corners indicating the scanning direction of the layer. The cell activations are again subsampled with a three by four window then fed to a feedforward layer of six tanh units. The feature maps corresponding to the feedforward unit activations are shown. The scanning and subsampling process is repeated until the feature maps corresponding to the cell activations of the uppermost MDLSTM layer are combined and collapsed to a single one-dimensional sequence of size 200 vectors, which is transcribed by a CTC layer containing 121 units. In this case all characters are correctly labelled except the second last one.

- A validation set was used for early stopping.
- The weights were randomly initialised from a Gaussian distribution with mean 0 and standard deviation 0.1.
- The inputs were standardised to have mean 0 and standard deviation 1 on the training set.
- 'White space' was trimmed from all input images using the colour of the top-left pixel.

### *Automatically Determined Parameters*

- The size of the input layer (determined by the input representation).
- The size of the output layer (determined by the target representation).
- The total number of weights in the hierarchy (determined by the layer sizes).
- The type of output layer (CTC or classification).
- The number of words in the dictionary used for CTC decoding. For all the tasks in this chapter, dictionary decoding was restricted to single words (Section 7.5.3.3). For dictionaries containing word variants the size was recorded as '# words / # variants'.

### *Hand-Tuned Parameters*

- The number of LSTM blocks in the recurrent layers in each level of the hierarchy (note that each layer in the same level always has the same number of blocks). If the layers contain $a$ blocks in the first level, $b$ blocks in the second level and $c$ blocks in the third layer, this will be abbreviated to 'Recurrent sizes: $a, b, c$.' Since the LSTM blocks always contain once cell, there will be a total a four hidden units per block for one-dimensional layers, and five per block for two-dimensional layers.
- The sizes of the two feedforward layers separating the hidden levels, abbreviated to 'Feedforward sizes: $a, b$'.
- The dimensions of the three subsampling windows, expressed as a list like $[w_1], [w_2], [w_3]$ for 1D networks and $[w_1, h_1], [w_2, h_2], [w_3, h_3]$ for 2D networks, where $w_i$ is the width of window $i$ and $h_i$ is the height. (Note that the distinction between width and height is somewhat arbitrary in general, but for the imaged-based experiments in this chapter it can be determined by the customary orientation of the image).
- The error measure used as the early-stopping criterion on the validation set. In most cases this will be the *label error rate* (Eqn. (2.18)).

## *9.2.1 Offline Arabic Handwriting Recognition*

The offline Arabic handwriting recognition competition at the 2009 International Conference on Document Analysis and Recognition (ICDAR 2009) (Märgner and Abed, 2009) was based on the publicly available IFN/ENIT database of handwritten Arabic words (Pechwitz et al., 2002). The data

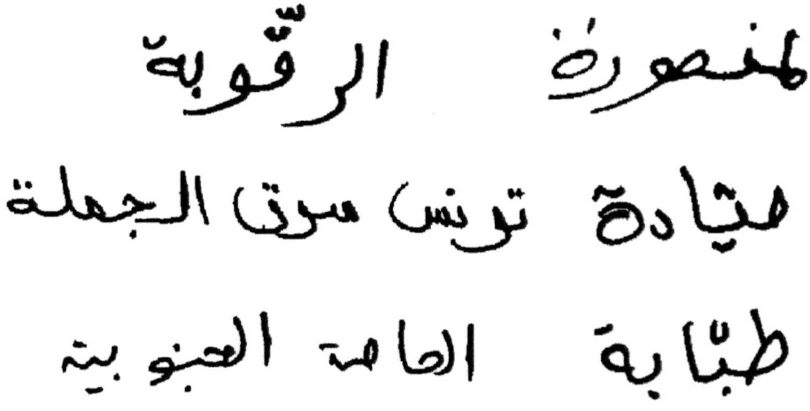

**Fig. 9.5 Offline Arabic word images**

consists of 32,492 black-and-white images of individual handwritten Tunisian town and village names, of which we used 30,000 for training, and 2,492 for validation. The images were extracted from forms filled in by over 400 Tunisian people. The forms were designed to simulate writing on a letter, and contained no lines or boxes to constrain the writing style. Example images are shown in Figure 9.5.

Each image was supplied with a manual transcription for the individual characters, and the postcode of the corresponding town. There were 120 distinct characters in total, including variant forms for initial, medial, final and isolated characters. The task was to identify the postcode, from a list of 937 town names and corresponding postcodes. Many of the town names had transcription variants, giving a total of 1,518 entries in the complete postcode lexicon.

The test data (which is not published) was divided into sets 'f' and 's'. The main competition results were based on set 'f'. Set 's' contains data collected in the United Arab Emirates using the same forms; its purpose was to test the robustness of the recognisers to regional writing variations. The entries were ranked according to their performance on set 'f'. In addition the recognition time of each of the systems was recorded on two extra subsets, labelled $t$ and $t_1$.

#### 9.2.1.1 Experimental Setup

Three HSRNNs were entered for the competition, with slightly different parameters. Within the competition, they were referred to as 'MDLSTM 9', 'MDLSTM 10' and 'MDLSTM 11'. The training parameters for the three systems are listed in Table 9.1. Networks 9 and 10 were identical, except that the label error rate was used for early stopping with the former, while the CTC error was used with the latter (in fact they were created during the same training run, with the weights recorded at different points). Network 11 had twice as many units in all the hidden layers as the other two networks (giving more than three times as many weights overall).

**Table 9.1 Networks entered for the offline Arabic handwriting competition at ICDAR 2009**

| Competition ID | MDLSTM 9 | MDLSTM 10 | MDLSTM 11 |
|---|---|---|---|
| Dimensions | 2 | 2 | 2 |
| Input Size | 1 | 1 | 1 |
| Output Size | 121 | 121 | 121 |
| Feedforward Sizes | 6, 20 | 6, 20 | 12, 40 |
| Recurrent Sizes | 2, 10, 50 | 2, 10, 50 | 4, 20, 100 |
| Windows | $[3,4]$, $[3,4]$, $[2,4]$ | $[3,4]$, $[3,4]$, $[2,4]$ | $[3,4]$, $[3,4]$, $[2,4]$ |
| Weights | 159,369 | 159,369 | 583,289 |
| Dictionary Size | 937 / 1,518 | 937 / 1,518 | 937 / 1,518 |
| Output Layer | CTC | CTC | CTC |
| Stopping Error | Label Error Rate | CTC loss | Label Error Rate |

#### 9.2.1.2 Results

The competition results are summarised in Table 9.2. The three HSRNNs (group ID 'MDLSTM') outperformed all other entries, in terms of both recognition rate and speed.

**Table 9.2 ICDAR 2009 offline Arabic handwriting recognition competition results** Results on test sets $f$ and $s$ are the percentage of correctly recognised postcodes. The average recognition time in ms per image on subset $t$ is shown in the last column. Best results shown in bold.

| Group-ID | System-ID | set $f$ (%) | set $s$ (%) | time (ms) |
|---|---|---|---|---|
| UOB-ENST | 1 | 82.07 | 69.99 | 812.69 |
| | 2 | 78.16 | 65.61 | 2365.48 |
| | 3 | 79.55 | 67.83 | 2236.58 |
| | 4 | 83.98 | 72.28 | 2154.48 |
| REGIM | 5 | 57.93 | 49.33 | 1564.75 |
| Ai2A | 6 | 85.58 | 70.44 | 1056,98 |
| | 7 | 82.21 | 66.45 | 519,61 |
| | 8 | 89.42 | 76.66 | 2583,64 |
| MDLSTM | 9 | 91.43 | 78.83 | 115.24 |
| | 10 | 91.37 | 78.89 | **114.61** |
| | 11 | **93.37** | **81.06** | 371.85 |
| RWTH-OCR | 12 | 85.51 | 71.33 | 17845.12 |
| | 13 | 85.69 | 72.54 | - |
| | 14 | 85.69 | 72.54 | - |
| | 15 | 83.90 | 65.99 | 542.12 |
| LITIS-MIRACL | 16 | 82.09 | 74.51 | 143269.81 |
| LSTS | 17 | 15.05 | 11.76 | 612.56 |

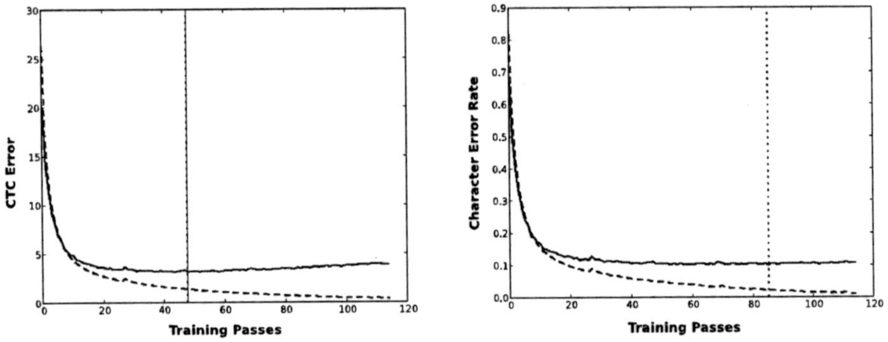


**Fig. 9.6 Error curves during training of networks 9 and 10.** The CTC error is shown on the left, and the character error is shown on the left. In both plots the solid line shows the error on the validation set, the dashed line shows the error on the training set, and the vertical dotted line indicates the point of lowest error on the validation set.

The overall difference in performance between networks 9 and 10 is negligible, suggesting that the choice of error measure used for early stopping is not crucial (although using the CTC loss for early stopping tend to lead to shorter training times). Of particular interest is that the performance on set $s$ (with handwriting from the United Arab Emirates) is about the same for both error measures. The original motivation for comparing the two stopping criteria was to see which would generalise better to test data drawn from a different distribution. One hypothesis, which was not supported by the experiment, was that using CTC loss as a stopping criterion would lead to less overfitting (because it is minimised sooner) and therefore better generalisation to different test data.

Network 11 gave about a 2% improvement n word recognition over networks 9 And 10. Although significant, this improvement comes at a cost of a more than threefold increase in word recognition time. For applications where time must be traded against accuracy, the number of units in the network layers (and hence the number of network weights) should be tuned accordingly.

Figure 9.6 shows the error curves for networks 9 and 10 during training. Note that, by the time the character error is minimised, the CTC error is already well past its minimum and has risen substantially. This is typical for CTC networks.

Network 9 finished training after 86 epochs, network 10 after 49 epochs and network 11 after 153 epochs.

## *9.2.2 Online Arabic Handwriting Recognition*

The online Arabic handwriting recognition competition at ICDAR 2009 (Abed et al., 2009) was based on the ADAB (**A**rabic **DA**ta**B**ase) database of Arabic online handwritten words. The database consists of 15,158 online pen traces, corresponding to handwritten Arabic words from 130 different writers. The pen traces are subdivided into individual strokes, with a list of $x$ and $y$ coordinates (recorded at a rate of 125 samples per second) provided for each stroke. The words were chosen from a dictionary of 984 Tunisian town and village names and manual transcriptions were provided along with the training set. Unlike the transcriptions provided for the offline competition, these were specified as unicode characters, with the distinction between medial, initial and final characters inferred from their context. There were therefore only 45 distinct characters in the transcriptions, as opposed to 120 for the offline competition in the previous section.

The training set was divided into three sets, with a fourth (unpublished) set was used for testing. The task was to correctly identify the town names in the test set, using the 984 word dictionary. The recognition rate with the top one, top five, and top ten answers was recorded by the organisers, and the entries were ranked according to their performance with one answer. The average recognition time per image on two subsets of the test set ('$t$' and '$t_1$') was also recorded. 1,523 of the training sequences were used as a validation set for early stopping.

#### 9.2.2.1 Experimental Setup

Two networks were submitted to the competition. For the first, 'online' network, the pen traces were fed directly into a one-dimensional HSRNN. This representation required three input units: two for the $x$ and $y$ pen coordinates, and one as an 'end-of-stroke' marker. For the second, 'offline' network, the pen traces were first transformed into a black-and-white image which was then processed by a two-dimensional HSRNN. The transformation was carried out by plotting straight lines between the pen coordinates in each of the strokes, and overlaying the resulting shapes to form an image. An illustration of the raw and offline representations is shown in Figure 9.7.

The network parameters are listed in Table 9.3. For both networks the same labels were used for the initial, medial and final forms of the characters, giving a total of 46 CTC output units. Better results would probably have been achieved by using different labels for different forms (as was done for the offline competition in Section 9.2.1).

#### 9.2.2.2 Results

In this competition the two HSRNNs were surpassed, in terms of both accuracy and speed, by the recognition systems entered by VisionObjects. The 'online' network took 85 epochs to train, while the 'offline' network took 91 epochs.

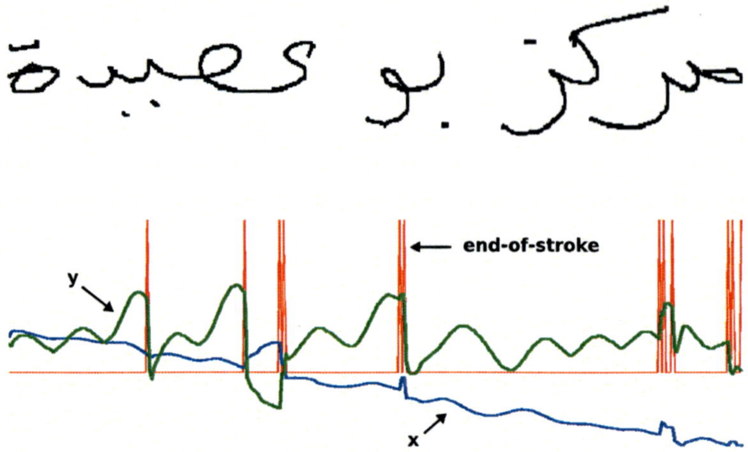


**Fig. 9.7 Online Arabic input sequences.** The offline image (above) was reconstructed from the online trace (below) by joining up the pen coordinates in the separate strokes.

**Table 9.3 Networks entered for the online Arabic handwriting competition at ICDAR 2009**

| Name | Online | Offline |
|---|---|---|
| Dimensions | 1 | 2 |
| Input Size | 3 | 1 |
| Output Size | 46 | 46 |
| Feedforward Sizes | 20, 60 | 8, 40 |
| Recurrent Sizes | 20, 60, 180 | 4, 20, 100 |
| Windows | [1], [2], [2] | [4, 3], [4, 2], [4, 2] |
| Weights | 423,926 | 550,334 |
| Dictionary Size | 984 | 984 |
| Output Layer | CTC | CTC |
| Stopping Error | Label Error Rate | Label Error Rate |

### *9.2.3 French Handwriting Recognition*

The French handwriting recognition competition at ICDAR 2009 (Grosicki and Abed, 2009) was based on a subset of the RIMES database (Grosicki et al., 2009) of French mail snippets. 44,195 hand-transcribed, isolated word images were provided for training, with a further 7542 images to be used as a validation set. Example images are shown in Figure 9.8. The transcriptions contained

**Table 9.4 ICDAR 2009 online Arabic handwriting recognition competition results** Accuracy is % correctly recognised images on dataset 4. The average recognition time in ms per image on subset *t*. Best results shown in bold.

| System | Accuracy (%) | time (ms) |
|---|---|---|
| HSRNN (online) | 95.70 | 1377.22 |
| HSRNN (offline) | 95.70 | 1712.45 |
| VisionObjects-1 | **98.99** | 172.67 |
| VisionObjects-2 | 98.99 | **69.41** |
| REGIM-HTK | 52.67 | 6402.24 |
| REGIM-CV | 13.99 | 7251.65 |
| REGIM-CV | 38.71 | 3571.25 |

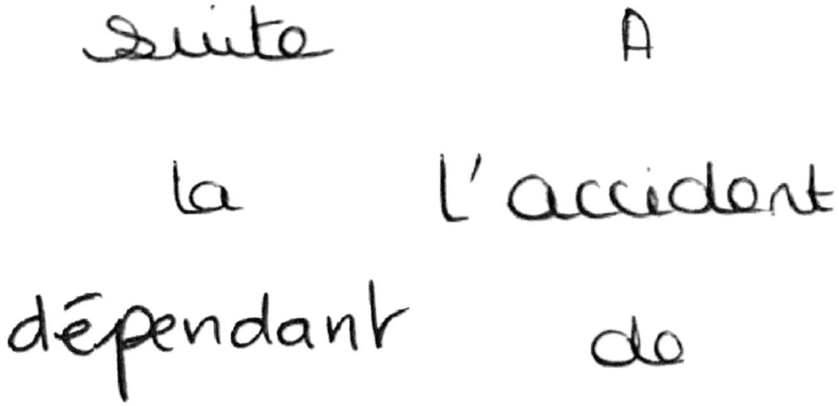


**Fig. 9.8 French word images**

81 distinct characters, including upper and lower case letters (with and without accents), numbers and punctuation. An additional (unpublished) test set of 7464 isolated words was used to assess performance.

Three tasks were defined on the test set, differing only in the dictionaries used for decoding. The first, referred to as 'WR1', defined a different 100 word dictionary for each word in the test set. Each dictionary contained the correct word plus 99 others randomly chosen from the test set. The second task ('WR2') used a 1,612 word dictionary, composed of all words in the test set. The third ('WR3') used a 5,534 word dictionary composed of all words in the test and training sets. For any given word in the test set, the 'WR1' dictionary was therefore a subset of the 'WR2' dictionary, which was a subset of the 'WR3' dictionary. Since adding incorrect words to a lexicon can only make decoding harder, the difficulty of the tasks strictly increased from 'WR1' to 'WR2' to 'WR3'.

**Table 9.5 Network entered for the French handwriting competition at ICDAR 2009**

| | |
|---|---|
| Dimensions | 2 |
| Input Size | 1 |
| Output Size | 82 |
| Feedforward Sizes | 6, 30 |
| Recurrent Sizes | 4, 20, 100 |
| Windows | $[2,3]$, $[2,3]$, $[2,3]$ |
| Weights | 531,842 |
| Dictionary Size (WR1) | 100 |
| Dictionary Size (WR2) | 1,612 |
| Dictionary Size (WR3) | 5,534 |
| Output Layer | CTC |
| Stopping Error | Label Error Rate |

#### 9.2.3.1 Experimental Setup

A single, two-dimensional HSRNN was submitted to the competition, with three different dictionaries (corresponding to the three tasks 'WR1', 'WR2' and 'WR3') used for decoding. Note that the network was only trained once, since the dictionary does not influence the CTC loss function. The network parameters are listed in Table 9.5.

#### 9.2.3.2 Results

The competition results for the three tasks are summarised in Table 9.6. The HSRNN performed best on all of them and won the competition. It was trained for 66 epochs.

### *9.2.4 Farsi/Arabic Character Classification*

The Farsi/Arabic character classification competition at ICDAR 2009 was based on data drawn from the Holda (Khosravi and Kabir, 2007), Farsi CENPARMI (Solimanpour et al., 2006) and Extended IFHCDB (Mozaffari et al., 2006) databases. The competition was divided into two tasks: letter classification and digit classification. In both cases a training set of hand-labelled images of isolated characters was provided to the competitors while an (unpublished) test set was used by the organisers to evaluate the systems. There were 34 distinct letters and 12 distinct digits. The training set for the letter task contained $106,181$ images, of which $10,619$ were used as a validation set, while the test set contained $107,992$ images. The training set for the digit task contained $106,000$ images, of which $3,008$ were used as a validation set, while the set contained $48,000$ images. Examples of both digit and letter images are shown in Figure 9.9.

**Table 9.6 ICDAR 2009 French handwriting recognition competition results.** Results are the percentage of correctly recognised images on the test set, using the dictionaries corresponding to tasks 'WR1', 'WR2' and 'WR3'. Best results are shown in bold.

| System | WR1 (%) | WR2 (%) | WR3 (%) |
|---|---|---|---|
| HSRNN | **98.4** | **93.2** | **91.0** |
| UPV | 96.5 | 86.1 | 83.2 |
| ParisTech(1) | 86.4 | 80.2 | 76.3 |
| IRISA | 91.2 | 79.6 | 74.7 |
| SIEMENS | 95.3 | 81.3 | 73.2 |
| ParisTech(2) | 82.1 | 72.4 | 68.0 |
| LITIS | 92.4 | 74.1 | 66.7 |
| ParisTech(3) | 79.9 | 63.8 | 58.7 |
| LSIS | - | - | 52.37 |
| ITESOFT | 74.6 | 59.4 | 50.44 |

#### 9.2.4.1 Experimental Setup

The images were preprocessed by padding them equally with white pixels on all sides to have a minimum width of 77 pixels and a minimum height of 95 pixels; this was found to improve performance for very small characters. The parameters for the networks used for the letter and digit classification competitions are listed in Table 9.7. Since these were *sequence classification tasks* (Section 2.3.1) a classification output layer was used instead of a CTC layer.

#### 9.2.4.2 Results

The competition results are summarised in Table 9.8. The HSRNN had the highest classification accuracy for the letter dataset and was pronounced the winner of the competition. The 'letter' network required 76 training epochs, while the 'digit' network required 44.

### *9.2.5 Phoneme Recognition*

This section compares the phoneme recognition accuracy of two HSRNNs and one non-hierarchical RNN, each with different input representations. The networks were evaluated on the TIMIT speech corpus (Garofolo et al., 1993), using the core test set and the reduced alphabet of 39 phonemes described in Section 7.6.2. The three input representations were:

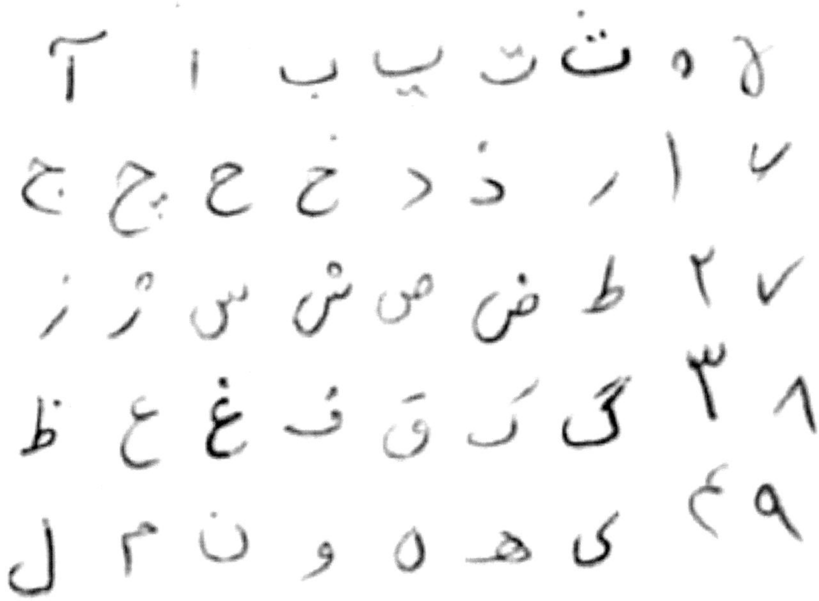

**Fig. 9.9 Farsi character images**

- Raw 16KHz sample sequences direct from the audio files.
- Spectrogram images.
- Mel-frequency cepstrum (MFC) coefficients.

The spectrograms were calculated from the sample sequences using the 'specgram' function of the 'matplotlib' python toolkit (Tosi, 2009), based on Welch's 'Periodogram' algorithm (Welch, 1967), with the following parameters: The Fourier transform windows were 254 samples wide with an overlap of 127 samples (corresponding to 15.875ms and 7.9375ms respectively). The MFC coefficients were calculated exactly as in Section 7.6.2. Figure 9.10 shows an example of the three representations for a single utterance from the TIMIT database.

### 9.2.5.1 Experimental Setup

The parameters for the three networks, referred to as 'raw', 'spectrogram' and 'MFC', are listed in 9.9. All three networks were evaluated with and without weight noise (Section 3.3.2.3) with a standard deviation of 0.075. The raw network has a single input because the TIMIT audio files are 'mono' and therefore have one channel per sample. Prefix search CTC decoding (Section 7.5) was used for all experiments, with a probability threshold of 0.995.

**Table 9.7 Networks entered for the Farsi/Arabic character classification competition at ICDAR 2009**

| Name | Letter | Digit |
|---|---|---|
| Dimensions | 2 | 2 |
| Input Size | 1 | 1 |
| Output Size | 34 | 12 |
| Feedforward Sizes | 9, 45 | 9, 45 |
| Recurrent Sizes | 4, 20, 100 | 4, 20, 100 |
| Windows | $[3,3]$, $[3,3]$, $[3,3]$ | $[3,3]$, $[3,3]$, $[3,3]$ |
| Weights | 562,754 | 553,932 |
| Output Layer | Classification | Classification |
| Stopping Error | Misclassification Rate | Misclassification Rate |

**Table 9.8 ICDAR 2009 Farsi/Arabic character classification competition results** Results are the classification error rates in the letter and digit test sets. Best results shown in bold.

| System | Letter Error (%) | Digit Error (%) |
|---|---|---|
| CEMATER-JU | 8.7 | 4.6 |
| HSRNN | **8.1** | 5.1 |
| REGIM | 11.7 | 5.7 |
| ECA | 10.5 | **4.1** |

**Table 9.9 Networks for phoneme recognition on TIMIT**

| Name | Raw | Spectrogram | MFC |
|---|---|---|---|
| Dimensions | 1 | 2 | 1 |
| Input Size | 1 | 1 | 39 |
| Output Size | 40 | 40 | 40 |
| Feedforward Sizes | 20, 40 | 6, 20 | - |
| Recurrent Sizes | 20, 40, 80 | 2, 10, 50 | 128 |
| Windows | $[6]$, $[6]$, $[6]$ | $[2,4]$, $[2,4]$, $[1,4]$, | - |
| Weights | 132,560 | 139,536 | 183,080 |
| Output Layer | CTC | CTC | CTC |
| Stopping Error | Label Error Rate | Label Error Rate | Label Error Rate |

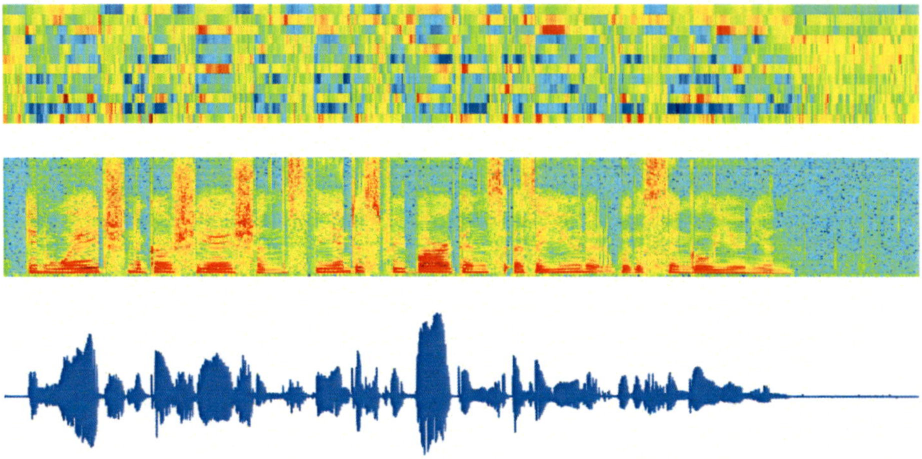

"In wage negotiations the industry bargains as a unit with a single union."

**Fig. 9.10 Three representations of a TIMIT utterance.** Both the MFC coefficients (top) and the spectrogram (middle) were calculated from the raw sequence of audio samples (bottom). Note the lower resolution and greater vertical and horizontal decorrelation of the MFC coefficients compared to the spectrogram.

#### 9.2.5.2 Results

The results of the experiments are presented in Table 9.10. Unlike the experiments in Section 7.6.1, repeated runs were not performed and it is therefore hard to determine if the differences are significant. However the 'spectrogram' network appears to give the best performance, with the 'raw' and 'MFC' networks approximately equal.

The number of training epochs was much lower for the MFC networks than either of the others; this echoes the results in Section 7.6.4, where learning from preprocessed online handwriting was found to be much faster (but not much more accurate) than learning from raw pen trajectories.

**Table 9.10 Phoneme recognition results on TIMIT.** The error measure is the phoneme error rate.

| Representation | Weight Noise | Error (%) | Epochs |
|---|---|---|---|
| Raw | ✗ | 30.5 | 79 |
| Raw | ✔ | 28.1 | 254 |
| MFC | ✗ | 29.5 | 27 |
| MFC | ✔ | 28.1 | 67 |
| Spectrogram | ✗ | 27.2 | 63 |
| Spectrogram | ✔ | 25.5 | 222 |

For the MFC network, training with input, rather than weight noise gives considerably better performance, as can be seen from Table 7.3. However Gaussian input noise does not help performance for the other two representations, because, as discussed in Section 3.3.2.2, it does not reflect the true variations in the input data. Weight noise, on the other hand, appears to be equally effective for all input representations.

# References


Abed, H.E., Margner, V., Kherallah, M., Alimi, A.M.: Online Arabic Handwriting Recognition Competition. In: 10th International Conference on Document Analysis and Recognition, ICDAR 2009, pp. 1388–1392. IEEE Computer Society (2009)

An, G.: The Effects of Adding Noise During Backpropagation Training on a Generalization Performance. Neural Computation 8(3), 643–674 (1996) ISSN 0899-7667

Bakker, B.: Reinforcement Learning with Long Short-Term Memory. In: Advances in Neural Information Processing Systems, vol. 14 (2002)

Baldi, P., Pollastri, G.: The Principled Design of Large-scale Recursive Neural Network Architectures–DAG-RNNs and the Protein Structure Prediction Problem. The Journal of Machine Learning Research 4, 575–602 (2003) ISSN 1533-7928

Baldi, P., Brunak, S., Frasconi, P., Soda, G., Pollastri, G.: Exploiting the Past and the Future in Protein Secondary Structure Prediction. Bioinformatics 15 (1999)

Baldi, P., Brunak, S., Frasconi, P., Pollastri, G., Soda, G.: Bidirectional Dynamics for Protein Secondary Structure Prediction. In: Sun, R., Giles, C.L. (eds.) Sequence Learning. LNCS (LNAI), vol. 1828, pp. 80–104. Springer, Heidelberg (2001)

Bayer, J., Wierstra, D., Togelius, J., Schmidhuber, J.: Evolving Memory Cell Structures for Sequence Learning. In: Alippi, C., Polycarpou, M., Panayiotou, C., Ellinas, G. (eds.) ICANN 2009. LNCS, vol. 5769, pp. 755–764. Springer, Heidelberg (2009)

Bengio, Y.: A Connectionist Approach to Speech Recognition. International Journal on Pattern Recognition and Artificial Intelligence 7(4), 647–668 (1993)

Bengio, Y.: Markovian Models for Sequential Data. Neural Computing Surveys 2, 129–162 (1999)

Bengio, Y., LeCun, Y.: Scaling learning algorithms towards AI. In: Bottou, L., Chapelle, O., DeCoste, D., Weston, J. (eds.) Large-Scale Kernel Machines. MIT Press (2007)

Bengio, Y., De Mori, R., Flammia, G., Kompe, R.: Global Optimization of a Neural Network–Hidden Markov Model Hybrid. IEEE Transactions on Neural Networks 3(2), 252–259 (1992)

Bengio, Y., Simard, P., Frasconi, P.: Learning Long-Term Dependencies with Gradient Descent is Difficult. IEEE Transactions on Neural Networks 5(2), 157–166 (1994)

Bengio, Y., LeCun, Y., Nohl, C., Burges, C.: LeRec: A NN/HMM Hybrid for On-line Handwriting Recognition. Neural Computation 7(6), 1289–1303 (1995)

Bengio, Y., Lamblin, P., Popovici, D., Larochelle, H.: Greedy Layer-wise Training of Deep Networks. In: Schölkopf, B., Platt, J., Hoffman, T. (eds.) Advances in Neural Information Processing Systems, vol. 19, pp. 153–160. MIT Press, Cambridge (2007)

Beringer, N.: Human Language Acquisition in a Machine Learning Task. In: International Conference on Spoken Language Processing (2004)

Bertolami, R., Bunke, H.: Multiple Classifier Methods for Offline Handwritten Text Line Recognition. In: 7th International Workshop on Multiple Classifier Systems, Prague, Czech Republic (2007)

Bishop, C.M.: Neural Networks for Pattern Recognition. Oxford University Press (1995)

Bishop, C.M.: Pattern Recognition and Machine Learning. Springer, Heidelberg (2006)

Bottou, L., LeCun, Y.: Graph Transformer Networks for Image Recognition. In: Proceedings of ISI (2005)

Bourlard, H., Morgan, N.: Connnectionist Speech Recognition: A Hybrid Approach. Kluwer Academic Publishers (1994)

Bourlard, H., Konig, Y., Morgan, N., Ris, C.: A new training algorithm for hybrid HMM/ANN speech recognition systems. In: 8th European Signal Processing Conference, vol. 1, pp. 101–104 (1996)

Bridle, J.S.: Probabilistic Interpretation of Feedforward Classification Network Outputs, with Relationships to Statistical Pattern Recognition. In: Fogleman-Soulie, F., Herault, J. (eds.) Neurocomputing: Algorithms, Architectures and Applications, pp. 227–236. Springer, Heidelberg (1990)

Broomhead, D., Lowe, D.: Multivariate Functional Interpolation and Adaptive Networks. Complex Systems 2, 321–355 (1988)

Byrd, R.H., Lu, P., Nocedal, J., Zhu, C.Y.: A Limited Memory Algorithm for Bound Constrained Optimization. SIAM Journal on Scientific Computing 16(6), 1190–1208 (1995)

Chang, J.: Near-Miss Modeling: A Segment-Based Approach to Speech Recognition. PhD thesis, Department of Electrical Engineering and Computer Science, Massachusetts Institute of Technology (1998)

Chen, J., Chaudhari, N.: Protein Secondary Structure Prediction with bidirectional LSTM networks. In: International Joint Conference on Neural Networks: Post-Conference Workshop on Computational Intelligence Approaches for the Analysis of Bio-data (CI-BIO) (August 2005)

Chen, J., Chaudhari, N.S.: Capturing Long-term Dependencies for Protein Secondary Structure Prediction. In: Yin, F.-L., Wang, J., Guo, C. (eds.) ISNN 2004, Part II. LNCS, vol. 3174, pp. 494–500. Springer, Heidelberg (2004)

Chen, R., Jamieson, L.: Experiments on the Implementation of Recurrent Neural Networks for Speech Phone Recognition. In: Proceedings of the Thirtieth Annual Asilomar Conference on Signals, Systems and Computers, pp. 779–782 (1996)

Decoste, D., Schölkopf, B.: Training Invariant Support Vector Machines. Machine Learning 46(1-3), 161–190 (2002)

Duda, R.O., Hart, P.E., Stork, D.G.: Pattern Classification. Wiley-Interscience Publication (2000)

Eck, D., Schmidhuber, J.: Finding Temporal Structure in Music: Blues Improvisation with LSTM Recurrent Networks. In: Bourlard, H. (ed.) Proceedings of the 2002 IEEE Workshop on Neural Networks for Signal Processing XII, pp. 747–756. IEEE, New York (2002)

Elman, J.L.: Finding Structure in Time. Cognitive Science 14, 179–211 (1990)

Fahlman, S.: Faster Learning Variations on Back-propagation: An Empirical Study. In: Touretzky, D., Hinton, G., Sejnowski, T. (eds.) Proceedings of the 1988 Connectionist Models Summer School, pp. 38–51. Morgan Kaufmann (1989)

Fernández, S., Graves, A., Schmidhuber, J.: An Application of Recurrent Neural Networks to Discriminative Keyword Spotting. In: de Sá, J.M., Alexandre, L.A., Duch, W., Mandic, D.P. (eds.) ICANN 2007. LNCS, vol. 4669, pp. 220–229. Springer, Heidelberg (2007)

Fernandez, S., Graves, A., Schmidhuber, J.: Phoneme Recognition in TIMIT with BLSTM-CTC. Technical Report IDSIA-04-08, IDSIA (April 2008)

Frasconi, P., Gori, M., Sperduti, A.: A General Framework for Adaptive Processing of Data Structures. IEEE Transactions on Neural Networks 9, 768–786 (1998)

Fukada, T., Schuster, M., Sagisaka, Y.: Phoneme Boundary Estimation Using Bidirectional Recurrent Neural Networks and its Applications. Systems and Computers in Japan 30(4), 20–30 (1999)

Garofolo, J.S., Lamel, L.F., Fisher, W.M., Fiscus, J.G., Pallett, D.S., Dahlgren, N.L.: DARPA TIMIT Acoustic Phonetic Continuous Speech Corpus CDROM (1993)

Gers, F.: Long Short-Term Memory in Recurrent Neural Networks. PhD thesis, Ecole Polytechnique Fédérale de Lausanne (2001)

Gers, F., Schraudolph, N., Schmidhuber, J.: Learning Precise Timing with LSTM Recurrent Networks. Journal of Machine Learning Research 3, 115–143 (2002)

Gers, F.A., Schmidhuber, J.: LSTM Recurrent Networks Learn Simple Context Free and Context Sensitive Languages. IEEE Transactions on Neural Networks 12(6), 1333–1340 (2001)

Gers, F.A., Schmidhuber, J., Cummins, F.: Learning to Forget: Continual Prediction with LSTM. Neural Computation 12(10), 2451–2471 (2000)

Giraud-Carrier, C., Vilalta, R., Brazdil, P.: Introduction to the Special Issue on Meta-Learning. Machine Learning 54(3), 187–193 (2004)

Glass, J.R.: A Probabilistic Framework for Segment-based Speech Recognition. Computer Speech and Language 17, 137–152 (2003)

Goller, C.: A Connectionist Approach for Learning Search-Control Heuristics for Automated Deduction Systems. PhD thesis, Fakultät für Informatik der Technischen Universität München (1997)

Graves, A., Schmidhuber, J.: Framewise Phoneme Classification with Bidirectional LSTM Networks. In: Proceedings of the 2005 International Joint Conference on Neural Networks (2005a)

Graves, A., Schmidhuber, J.: Framewise Phoneme Classification with Bidirectional LSTM and Other Neural Network Architectures. Neural Networks 18(5-6), 602–610 (2005b)

Graves, A., Schmidhuber, J.: Offline Handwriting Recognition with Multidimensional Recurrent Neural Networks. In: Koller, D., Schuurmans, D., Bengio, Y., Bottou, L. (eds.) Advances in Neural Information Processing Systems, vol. 21, pp. 545–552. MIT Press (2009)

Graves, A., Beringer, N., Schmidhuber, J.: Rapid Retraining on Speech Data with LSTM Recurrent Networks. Technical Report IDSIA-09-05, IDSIA (2005a)

Graves, A., Fernández, S., Schmidhuber, J.: Bidirectional LSTM Networks for Improved Phoneme Classification and Recognition. In: Duch, W., Kacprzyk, J., Oja, E., Zadrożny, S. (eds.) ICANN 2005. LNCS, vol. 3697, pp. 799–804. Springer, Heidelberg (2005)

Graves, A., Fernández, S., Gomez, F., Schmidhuber, J.: Connectionist Temporal Classification: Labelling Unsegmented Sequence Data with Recurrent Neural Networks. In: Proceedings of the International Conference on Machine Learning, ICML 2006, Pittsburgh, USA (2006)

Graves, A., Fernández, S., Schmidhuber, J.: Multi-dimensional Recurrent Neural Networks. In: de Sá, J.M., Alexandre, L.A., Duch, W., Mandic, D.P. (eds.) ICANN 2007. LNCS, vol. 4668, pp. 549–558. Springer, Heidelberg (2007)

Graves, A., Fernández, S., Liwicki, M., Bunke, H., Schmidhuber, J.: Unconstrained Online Handwriting Recognition with Recurrent Neural Networks. In: Platt, J., Koller, D., Singer, Y., Roweis, S. (eds.) Advances in Neural Information Processing Systems, vol. 20. MIT Press, Cambridge (2008)

Graves, A., Liwicki, M., Fernández, S., Bertolami, R., Bunke, H., Schmidhuber, J.: A Novel Connectionist System for Unconstrained Handwriting Recognition. IEEE Transactions on Pattern Analysis and Machine Intelligence 31(5), 855–868 (2009)

Grosicki, E., Abed, H.E.: ICDAR 2009 Handwriting Recognition Competition. In: 10th International Conference on Document Analysis and Recognition, pp. 1398–1402 (2009)

Grosicki, E., Carre, M., Brodin, J.-M., Geoffrois, E.: Results of the RIMES Evaluation Campaign for Handwritten Mail Processing. In: International Conference on Document Analysis and Recognition, pp. 941–945 (2009)

Halberstadt, A.K.: Heterogeneous Acoustic Measurements and Multiple Classifiers for Speech Recognition. PhD thesis, Department of Electrical Engineering and Computer Science, Massachusetts Institute of Technology (1998)

Hammer, B.: On the Approximation Capability of Recurrent Neural Networks. Neurocomputing 31(1–4), 107–123 (2000)

Hammer, B.: Recurrent Networks for Structured Data - a Unifying Approach and Its Properties. Cognitive Systems Research 3, 145–165 (2002)

Hennebert, J., Ris, C., Bourlard, H., Renals, S., Morgan, N.: Estimation of Global Posteriors and Forward-backward Training of Hybrid HMM/ANN Systems. In: Proc. of the European Conference on Speech Communication and Technology (Eurospeech 1997), pp. 1951–1954 (1997)

Hestenes, M.R., Stiefel, E.: Methods of Conjugate Gradients for Solving Linear Systems. Journal of Research of the National Bureau of Standards 49(6), 409–436 (1952)

Hifny, Y., Renals, S.: Speech Recognition using Augmented Conditional Random Fields. Trans. Audio, Speech and Lang. Proc. 17, 354–365 (2009)

Hinton, G.E., van Camp, D.: Keeping Neural Networks Simple by Minimizing the Description Length of the Weights. In: Conference on Learning Theory, pp. 5–13 (1993)

Hinton, G.E., Osindero, S., Teh, Y.-W.: A Fast Learning Algorithm for Deep Belief Nets. Neural Computation 18(7), 1527–1554 (2006)

Hochreiter, S.: Untersuchungen zu Dynamischen Neuronalen Netzen. PhD thesis, Institut für Informatik, Technische Universität München (1991)

Hochreiter, S., Schmidhuber, J.: Long Short-Term Memory. Neural Computation 9(8), 1735–1780 (1997)

Hochreiter, S., Bengio, Y., Frasconi, P., Schmidhuber, J.: Gradient Flow in Recurrent Nets: the Difficulty of Learning Long-term Dependencies. In: Kremer, S.C., Kolen, J.F. (eds.) A Field Guide to Dynamical Recurrent Neural Networks. IEEE Press (2001a)

Hochreiter, S., Bengio, Y., Frasconi, P., Schmidhuber, J.: Gradient flow in Recurrent Nets: the Difficulty of Learning Long-term Dependencies. In: Kremer, S.C., Kolen, J.F. (eds.) A Field Guide to Dynamical Recurrent Neural Networks. IEEE Press (2001b)

Hochreiter, S., Heusel, M., Obermayer, K.: Fast Model-based Protein Homology Detection without Alignment. Bioinformatics (2007)

Hopfield, J.J.: Neural Networks and Physical Systems with Emergent Collective Computational Abilities. Proceedings of the National Academy of Sciences of the United States of America 79(8), 2554–2558 (1982)

Hornik, K., Stinchcombe, M., White, H.: Multilayer Feedforward Networks are Universal Approximators. Neural Networks 2(5), 359–366 (1989)

Hülsken, F., Wallhoff, F., Rigoll, G.: Facial Expression Recognition with Pseudo-3D Hidden Markov Models. In: Radig, B., Florczyk, S. (eds.) DAGM 2001. LNCS, vol. 2191, pp. 291–297. Springer, Heidelberg (2001)

Jaeger, H.: The "Echo State" Approach to Analysing and Training Recurrent Neural Networks. Technical Report GMD Report 148, German National Research Center for Information Technology (2001)

Jim, K.-C., Giles, C., Horne, B.: An Analysis of Noise in Recurrent Neural Networks: Convergence and Generalization. IEEE Transactions on Neural Networks 7(6), 1424–1438 (1996)

Jiten, J., Mérialdo, B., Huet, B.: Multi-dimensional Dependency-tree Hidden Markov Models. In: International Conference on Acoustics, Speech, and Signal Processing (2006)

Johansson, S., Atwell, R., Garside, R., Leech, G.: The tagged LOB corpus user's manual; Norwegian Computing Centre for the Humanities (1986)

Johnson, M.T.: Capacity and Complexity of HMM Duration Modeling techniques. IEEE Signal Processing Letters 12(5), 407–410 (2005)

Jordan, M.I.: Attractor dynamics and parallelism in a connectionist sequential machine, pp. 112–127. IEEE Press (1990)

Joshi, D., Li, J., Wang, J.: Parameter Estimation of Multi-dimensional Hidden Markov Models: A Scalable Approach. In: Proc. of the IEEE International Conference on Image Processing (ICIP 2005), pp. 149–152 (2005)

Kadous, M.W.: Temporal Classification: Extending the Classification Paradigm to Multivariate Time Series. PhD thesis, School of Computer Science & Engineering, University of New South Wales (2002)

Kershaw, D., Robinson, A., Hochberg, M.: Context-Dependent Classes in a Hybrid Recurrent Network-HMM Speech Recognition System. In: Touretzky, D.S., Mozer, M.C., Hasselmo, M.E. (eds.) Advances in Neural Information Processing Systems, vol. 8, pp. 750–756. MIT Press (1996)

Khosravi, H., Kabir, E.: Introducing a Very Large Dataset of Handwritten Farsi Digits and a Study on their Varieties. Pattern Recogn. Lett. 28, 1133–1141 (2007)

Kohonen, T.: Self-organization and Associative Memory, 3rd edn. Springer, New York (1989)

Koistinen, P., Holmström, L.: Kernel Regression and Backpropagation Training with Noise. In: Moody, J.E., Hanson, S.J., Lippmann, R. (eds.) Advances in Neural Information Processing Systems, vol. 4, pp. 1033–1039. Morgan Kaufmann (1991)

Lafferty, J.D., McCallum, A., Pereira, F.C.N.: Conditional Random Fields: Probabilistic Models for Segmenting and Labeling Sequence Data. In: Proceedings of the Eighteenth International Conference on Machine Learning, ICML 2001, pp. 282–289. Morgan Kaufmann Publishers Inc. (2001)

Lamel, L., Gauvain, J.: High Performance Speaker-Independent Phone Recognition Using CDHMM. In: Proc. Eurospeech (September 1993)

Lang, K.J., Waibel, A.H., Hinton, G.E.: A Time-delay Neural Network Architecture for Isolated Word Recognition. Neural Networks 3(1), 23–43 (1990)

LeCun, Y., Bottou, L., Bengio, Y.: Reading Checks with Graph Transformer Networks. In: International Conference on Acoustics, Speech, and Signal Processing, vol. 1, pp. 151–154. IEEE (1997)

LeCun, Y., Bottou, L., Bengio, Y., Haffner, P.: Gradient-Based Learning Applied to Document Recognition. Proceedings of the IEEE 86(11), 2278–2324 (1998a)

LeCun, Y., Bottou, L., Bengio, Y., Haffner, P.: Gradient-based Learning Applied to Document Recognition. Proceedings of the IEEE, 1–46 (1998b)

LeCun, Y.A., Bottou, L., Orr, G.B., Müller, K.-R.: Efficient BackProp. In: Orr, G.B., Müller, K.-R. (eds.) NIPS-WS 1996. LNCS, vol. 1524, pp. 9–50. Springer, Heidelberg (1998)

Lee, K.-F., Hon, H.-W.: Speaker-independent Phone Recognition Using Hidden Markov Models. IEEE Transactions on Acoustics, Speech, and Signal Processing 37(11), 1641–1648 (1989)

Li, J., Najmi, A., Gray, R.M.: Image Classification by a Two-Dimensional Hidden Markov Model. IEEE Transactions on Signal Processing 48(2), 517–533 (2000)

Lin, T., Horne, B.G., Tiño, P., Giles, C.L.: Learning Long-Term Dependencies in NARX Recurrent Neural Networks. IEEE Transactions on Neural Networks 7(6), 1329–1338 (1996)

Lindblad, T., Kinser, J.M.: Image Processing Using Pulse-Coupled Neural Networks. Springer-Verlag New York, Inc. (2005)

Liwicki, M., Bunke, H.: Handwriting Recognition of Whiteboard Notes. In: Proc. 12th Conf. of the International Graphonomics Society, pp. 118–122 (2005a)

Liwicki, M., Bunke, H.: IAM-OnDB - an On-Line English Sentence Database Acquired from Handwritten Text on a Whiteboard. In: Proc. 8th Int. Conf. on Document Analysis and Recognition, vol. 2, pp. 956–961 (2005b)

Liwicki, M., Graves, A., Fernández, S., Bunke, H., Schmidhuber, J.: A Novel Approach to On-Line Handwriting Recognition Based on Bidirectional Long Short-Term Memory Networks. In: Proceedings of the 9th International Conference on Document Analysis and Recognition, ICDAR 2007 (September 2007)

MacKay, D.J.C.: Probable Networks and Plausible Predictions - a Review of Practical Bayesian Methods for Supervised Neural Networks. Network: Computation in Neural Systems 6, 469–505 (1995)

Märgner, V., Abed, H.E.: ICDAR 2009 Arabic Handwriting Recognition Competition. In: 10th International Conference on Document Analysis and Recognition, pp. 1383–1387 (2009)

Marti, U.-V., Bunke, H.: Using a Statistical Language Model to Improve the Performance of an HMM-based Cursive Handwriting Recognition System. Int. Journal of Pattern Recognition and Artificial Intelligence 15, 65–90 (2001)

Marti, U.-V., Bunke, H.: The IAM Database: An English Sentence Database for Offline Handwriting Recognition. International Journal on Document Analysis and Recognition 5, 39–46 (2002)

McCarter, G., Storkey, A.: Air Freight Image Segmentation Database (2007)

McCulloch, W.S., Pitts, W.: A Logical Calculus of the Ideas Immanent in Nervous Activity, pp. 15–27. MIT Press (1988)

Ming, J., Smith, F.J.: Improved Phone Recognition Using Bayesian Triphone Models. In: ICASSP, vol. 1, pp. 409–412 (1998)

Mohamed, A., Dahl, G., Hinton, G.: Acoustic Modeling using Deep Belief Networks. IEEE Transactions on Audio, Speech, and Language Processing (99) (2011)

Morris, J., Lussier, E.F.: Combining Phonetic Attributes Using Conditional Random Fields. In: Proc. Interspeech 2006 (2006)

Mozaffari, S., Faez, K., Faradji, F., Ziaratban, M., Golzan, S.M.: Comprehensive Isolated Farsi/Arabic Character Database for Handwritten OCR Research. In: Lorette, G. (ed.) Tenth International Workshop on Frontiers in Handwriting Recognition. Suvisoft (2006)

Mozer, M.C.: Induction of Multiscale Temporal Structure. In: Moody, J.E., Hanson, S.J., Lippmann, R.P. (eds.) Advances in Neural Information Processing Systems, vol. 4, pp. 275–282. Morgan Kaufmann Publishers (1992)

Murray, A.F., Edwards, P.J.: Enhanced MLP Performance and Fault Tolerance Resulting from Synaptic Weight Noise During Training. IEEE Transactions on Neural Networks 5, 792–802 (1994)

Navarro, G.: A guided tour to approximate string matching. ACM Computing Surveys 33(1), 31–88 (2001)

Neal, R.M.: Bayesian Learning for Neural Networks. Springer, New York (1996)

Neto, J., Almeida, L., Hochberg, M., Martins, C., Nunes, L., Renals, S., Robinson, A.: Speaker Adaptation for Hybrid HMM-ANN Continuous Speech Recognition System. In: Proceedings of Eurospeech 1995, vol. 1, pp. 2171–2174 (1995)

Pang, X., Werbos, P.J.: Neural Network Design for J Function Approximation in Dynamic Programming. Mathematical Modeling and Scientific Computing 5(2/3) (1996)

Pechwitz, M., Maddouri, S.S., Mrgner, V., Ellouze, N., Amiri, H.: IFN/ENIT - Database of Handwritten Arabic Words. In: Colloque International Francophone sur l'Écrit et le Document, pp. 129–136 (2002)

Plate, T.A.: Holographic Recurrent Networks. In: Giles, C.L., Hanson, S.J., Cowan, J.D. (eds.) Advances in Neural Information Processing Systems, vol. 5, pp. 34–41. Morgan Kaufmann (1993)

Plaut, D.C., Nowlan, S.J., Hinton, G.E.: Experiments on Learning by Back-Propagation. Technical Report CMU–CS–86–126, Carnegie–Mellon University (1986)

Pollastri, G., Vullo, A., Frasconi, P., Baldi, P.: Modular DAG-RNN Architectures for Assembling Coarse Protein Structures. Journal of Computational Biology 13(3), 631–650 (2006)

Rabiner, L.R.: A Tutorial on Hidden Markov Models and Selected Applications in Speech Recognition. Proc. IEEE 77(2), 257–286 (1989)

Reisenhuber, M., Poggio, T.: Hierarchical Models of Object Recognition in Cortex. Nature Neuroscience 2(11), 1019–1025 (1999)

Renals, S., Morgan, N., Bourlard, H., Cohen, M., Franco, H.: Connectionist Probability Estimators in HMM Speech Recognition. IEEE Transactions Speech and Audio Processing (1993)

Riedmiller, M., Braun, H.: A Direct Adaptive Method for Faster Backpropagation Learning: The RPROP algorithm. In: Proc. of the IEEE Intl. Conf. on Neural Networks, San Francisco, CA, pp. 586–591 (1993)

Robinson, A., Holdsworth, J., Patterson, J., Fallside, F.: A comparison of preprocessors for the cambridge recurrent error propagation network speech recognition system. In: Proceedings of the First International Conference on Spoken Language Processing, ICSLP 1990 (1990)

Robinson, A.J.: Several Improvements to a Recurrent Error Propagation Network Phone Recognition System. Technical Report CUED/F-INFENG/TR82, University of Cambridge (1991)

Robinson, A.J.: An Application of Recurrent Nets to Phone Probability Estimation. IEEE Transactions on Neural Networks 5(2), 298–305 (1994)

Robinson, A.J., Fallside, F.: The Utility Driven Dynamic Error Propagation Network. Technical Report CUED/F-INFENG/TR.1, Cambridge University Engineering Department (1987)

Robinson, A.J., Almeida, L., Boite, J.-M., Bourlard, H., Fallside, F., Hochberg, M., Kershaw, D., Kohn, P., Konig, Y., Morgan, N., Neto, J.P., Renals, S., Saerens, M., Wooters, C.: A Neural Network Based, Speaker Independent, Large Vocabulary, Continuous Speech Recognition System: the Wernicke Project. In: Proc. of the Third European Conference on Speech Communication and Technology (Eurospeech 1993), pp. 1941–1944 (1993)

Rosenblatt, F.: The Perceptron: a Probabilistic Model for Information Storage and Organization in the Brain. Psychological Review 65, 386–408 (1958)

Rosenblatt, F.: Principles of Neurodynamics. Spartan, New York (1963)

Rumelhart, D.E., Hinton, G.E., Williams, R.J.: Learning Internal Representations by Error Propagation, pp. 318–362. MIT Press (1986)

Russell, S., Norvig, P.: Artificial Intelligence: A Modern Approach, 2nd edn. Prentice-Hall, Englewood Cliffs (2003)

Sainath, T., Ramabhadran, B., Picheny, M.: An Exploration of Large Vocabulary Tools for Small Vocabulary Phonetic Recognition. In: IEEE Workshop on Automatic Speech Recognition Understanding, ASRU 2009, pp. 359–364 (2009)

Schmidhuber, J.: Learning Complex Extended Sequences using the principle of history compression. Neural Computing, 234–242 (1992)

Schmidhuber, J., Wierstra, D., Gagliolo, M., Gomez, F.: Training Recurrent Networks by Evolino. Neural Computation 19(3), 757–779 (2007)

Schraudolph, N.: Fast Curvature Matrix-Vector Products for Second-Order Gradient Descent. Neural Computation 14(7), 1723–1738 (2002)

Schuster, M.: On Supervised Learning from Sequential Data With Applications for Speech Recognition. PhD thesis, Nara Institute of Science and Technology, Kyoto, Japan (1999)

Schuster, M., Paliwal, K.K.: Bidirectional Recurrent Neural Networks. IEEE Transactions on Signal Processing 45, 2673–2681 (1997)

Senior, A., Robinson, A.J.: Forward-Backward Retraining of Recurrent Neural Networks. In: Touretzky, D.S., Mozer, M.C., Hasselmo, M.E. (eds.) Advances in Neural Information Processing Systems, vol. 8, pp. 743–749. The MIT Press (1996)

Sha, F., Saul, L.K.: Large Margin Hidden Markov Models for Automatic Speech Recognition. In: Advances in Neural Information Processing Systems, pp. 1249–1256 (2006)

Shewchuk, J.R.: An Introduction to the Conjugate Gradient Method Without the Agonizing Pain. Technical report, Carnegie Mellon University, Pittsburgh, PA, USA (1994)

Simard, P.Y., Steinkraus, D., Platt, J.C.: Best Practices for Convolutional Neural Networks Applied to Visual Document Analysis. In: ICDAR 2003: Proceedings of the Seventh International Conference on Document Analysis and Recognition. IEEE Computer Society (2003)

Solimanpour, F., Sadri, J., Suen, C.Y.: Standard Databases for Recognition of Handwritten Digits, Numerical Strings, Legal Amounts, Letters and Dates in Farsi Language. In: Lorette, G. (ed.) Tenth International Workshop on Frontiers in Handwriting Recognition, Suvisoft (October 2006)

Sperduti, A., Starita, A.: Supervised Neural Networks for the Classification of Structures. IEEE Transactions on Neural Networks 8(3), 714–735 (1997)

Thireou, T., Reczko, M.: Bidirectional Long Short-Term Memory Networks for Predicting the Subcellular Localization of Eukaryotic Proteins. IEEE/ACM Trans. Comput. Biol. Bioinformatics 4(3), 441–446 (2007)

Tosi, S.: Matplotlib for Python Developers. Packt Publishing (2009)

Trentin, E., Gori, M.: Robust combination of neural networks and hidden Markov models for speech recognition. IEEE Transactions on Neural Networks 14(6), 1519–1531 (2003)

Vapnik, V.N.: The Nature of Statistical Learning Theory. Springer-Verlag New York, Inc. (1995)

Verbmobil. Database Version 2.3 (2004)

Welch, P.: The Use of Fast Fourier Transform for the Estimation of Power Spectra: a Method Based on Time Averaging over Short, Modified Periodograms. IEEE Transactions on Audio and Electroacoustics 15(2), 70–73 (1967)

Werbos, P.: Backpropagation Through Time: What It Does and How to Do It. Proceedings of the IEEE 78(10), 1550–1560 (1990)

Werbos, P.J.: Generalization of Backpropagation with Application to a Recurrent Gas Market Model. Neural Networks 1 (1988)

Wierstra, D., Gomez, F.J., Schmidhuber, J.: Modeling systems with internal state using evolino. In: GECCO 2005: Proceedings of the 2005 Conference on Genetic and Evolutionary Computation, pp. 1795–1802. ACM Press (2005)

Williams, R.J., Zipser, D.: Gradient-Based Learning Algorithms for Recurrent Networks and Their Computational Complexity. In: Chauvin, Y., Rumelhart, D.E. (eds.) Back-Propagation: Theory, Architectures and Applications, pp. 433–486. Lawrence Erlbaum Publishers (1995)

Wu, L., Baldi, P.: A Scalable Machine Learning Approach to Go. In: Schlkopf, B., Platt, J., Hoffman, T. (eds.) Advances in Neural Information Processing Systems, vol. 19, pp. 1521–1528. MIT Press (2006)

Young, S., Russell, N., Thornton, J.: Token Passing: A Simple Conceptual Model for Connected Speech Recognition Systems. Technical Report CUED/F-INFENG/TR38, Cambridge University Engineering Dept., Cambridge, UK (1989)

Young, S., Evermann, G., Gales, M., Hain, T., Kershaw, D., Liu, X., Moore, G., Odell, J., Ollason, D., Povey, D., Valtchev, V., Woodland, P.: The HTK Book. Cambridge University Engineering Department, HTK version 3.4 edition (December 2006)

Yu, D., Deng, L., Acero, A.: A Lattice Search Technique for a Long-Contextual-Span Hidden Trajectory Model of Speech. Speech Communication 48(9), 1214–1226 (2006)

Zavaliagkos, G., Austin, S., Makhoul, J., Schwartz, R.M.: A Hybrid Continuous Speech Recognition System Using Segmental Neural Nets with Hidden Markov Models. International Journal of Pattern Recognition and Artificial Intelligence 7(4), 949–963 (1993)

Zimmermann, H.G., Grothmann, R., Schaefer, A.M., Tietz, C.: Identification and Forecasting of Large Dynamical Systems by Dynamical Consistent Neural Networks. In: Haykin, S., Principe, J., Sejnowski, T., McWhirter, J. (eds.) New Directions in Statistical Signal Processing: From Systems to Brain, pp. 203–242. MIT Pres (2006a)

Zimmermann, M., Chappelier, J.-C., Bunke, H.: Offline Grammar-based Recognition of Handwritten Sentences. IEEE Transactions on Pattern Analysis and Machine Intelligence 28(5), 818–821 (2006b)

# Acknowledgements

I would first like to thank my supervisor Jürgen Schmidhuber for his guidance and support throughout the Ph.D. thesis on which this book was based. I would also like to thank my co-authors Santiago Fernández, Nicole Beringer, Faustino Gomez and Douglas Eck, and everyone else I collaborated with at IDSIA and the Technical University of Munich, for making the stimulating and creative places to work. Thanks to Tom Schaul for proofreading an early draft of the book, and to Marcus Hutter for his assistance during Chapter 7. I am grateful to Marcus Liwicki, Horst Bunke and Roman Bertolami for their expert collaboration on handwriting recognition. A special mention to all my friends in Lugano, Munich and elsewhere who made the whole thing worth doing: Frederick Ducatelle, Matteo Gagliolo, Nikos Mutsanas, Ola Svensson, Daniil Ryabko, John Paul Walsh, Adrian Taruttis, Andreas Brandmaier, Christian Osendorfer, Thomas Rückstieß, Justin Bayer, Murray Dick, Luke Williams, John Lord, Sam Mungall, David Larsen and all the rest. But most of all I would like to thank my family, my wife Alison and my children Liam and Nina for being there when I needed them most.

Alex Graves is a Junior Fellow of the Canadian Institute for Advanced Research.